PICTORIAL
ANATOMY
OF
THE CAT

"One would think that the conversing with dead and stinking Carkases (that are not onely hideous objects in themselves, but made more ghastly by the putting us in mind that our selves must be such) should be not onely a very melancholy, but a very hated employment. And yet . . . I confess its Instructiveness has not onely so reconciled me to it . . . that I have often spent hours much less delightfully, not onely in Courts, but even in Libraries, then in tracing in those forsaken Mansions, the inimitable Workmanship of the Omniscient Architect."

ROBERT BOYLE (1627-91)

PICTORIAL ANATOMY OF THE CAT

REVISED EDITION

ILLUSTRATIONS & TEXT BY STEPHEN G. GILBERT

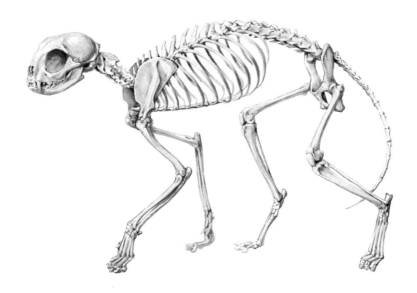

University of Washington Press   Seattle and London

Copyright © 1968 by the University of Washington Press
Thirteenth printing, 2000
Designed by Joseph Erceg and Stephen G. Gilbert
Printed and bound in Hong Kong

Library of Congress Catalog Card Number 67-21200
ISBN 0-295-95454-X

Published in Canada by the University of Toronto Press
in 1976, 1980, 1994, 2000.
ISBN 0-8020-2249-9

The paper used in this publication meets the minimum requirements of American National Standard for Information Sciences—Permanence of Paper for Printed Library Materials, ANSI Z39.48-1984. ⊗

PREFACE

This book is designed for use as a dissection guide in comparative vertebrate anatomy or in mammalian anatomy. The material covered and the time allotted to such courses varies considerably, and the illustrations are therefore designed to enable the instructor to point out the important features of areas which cannot be dissected in detail by every student.

In some instances I have departed from the traditional methods of dissection. It is customary, for instance, to bisect and retract superficial muscles rather than removing them completely, but my instructions call for the complete removal of the superficial muscles on the left side. This is necessary for adequate exposure of the deep muscles; the muscles on the right may be left intact or bisected and retracted to re-establish the original relations. The instructor should feel free to modify the dissecting instructions for the muscles and for any other areas to fit his needs.

Many variations occur in the structure of individual specimens. The text describes the norm, and for the sake of brevity the numerous variations which may be encountered in the laboratory are not described. Students should not be surprised, therefore, if their specimens do not exactly resemble the illustrations, and they should examine a number of other specimens beside their own. Variations will be found in the number of vertebrae, in the sizes and shapes of the muscles and of the viscera, and in the branching of nerves and vessels, particularly the veins.

I have used a combination of diagrammatic marginal illustrations, which are labeled directly, and larger realistic illustrations, which are labeled with numbered keys. The student should read each section through, following the text and studying the marginal diagrams before beginning the dissection of a given area. As he dissects and identifies various structures he may find it convenient to write the names of the structures directly on the illustrations, or to write them on tracing paper overlays which can be removed for purposes of review.

In the matter of terminology and detail I have followed Libbie Hyman's widely used *Comparative Vertebrate Anatomy*, departing from her treatment principally in that I have adapted my dissecting instructions and descriptive text to a pictorial approach. There are two schools of thought about the value of illustrations in teaching anatomy. Dr. Hyman cautions her students to "remember that the laboratory instructors are familiar with all of the figures in the various textbooks and that undue resemblance between your drawings and such figures will reflect upon your honesty and raise a suspicion that you have not been exerting yourself in the laboratory." On the use of illustrations she writes:

> I have always believed that zoology is best studied and learned not out of books but by actual experience with and handling of material. . . . I have avoided giving any illustrations of the animals under dissection and, in general, hope that the student will learn the story of the evolution of the various vertebrate organ systems by comparing their condition in the different animals dissected rather than by looking at diagrams in books.

The case for illustrations was well stated by the great Renaissance anatomist, Andreas Vesalius (1514-64). Before Vesalius, anatomy was taught by a professor who read from Galen while an assistant dissected the corpse. Neither the professor nor his students dirtied their hands with cadavers. The fact that the structures dissected by the assistant did not always correspond to Galen's text was ignored; anatomists had more faith in the books of Galen than they had in their own observations. Vesalius singlehandedly changed these methods of teaching. He made dissections himself and described what he saw instead of relying on the observations of Galen. In 1543 he published his beautifully illustrated *De Humani Corporis Fabrica (On the Structure of the Human Body),* and with it did more than any other anatomist to change our understanding of anatomy and to revolutionize teaching methods. His students did not look at his illustrations instead of at the cadaver; on the contrary, they were able, for the first time, to study the cadaver intelligently because the illustrations showed them what to look for. It is interesting, however, to find that Vesalius' jealous contemporaries attacked him with the arguments still being repeated, over four hundred years later, by Hyman. Vesalius answered them with the following words, from the preface of *De Humani Corporis Fabrica:*

> But here there comes into my mind the judgement of certain men who vehemently condemn the practice of setting before the eyes of students, as we do with parts of plants, delineations, be they never so accurate, of the parts of the human body. These, they say, ought to be learned not by pictures but by careful dissection and examination of the things themselves. As if, forsooth, my object in adding to the text of my discourse images of the parts, which are most faithful, and which I wish could be free from the risk of being spoiled by the printers, was that students should rely upon them and refrain from dissecting bodies; whereas my practice has rather been to encourage students of medicine in every way I could to perform dissections with their own hands.

STEPHEN G. GILBERT

February, 1968

CONTENTS

PICTORIAL
ANATOMY
OF
THE CAT

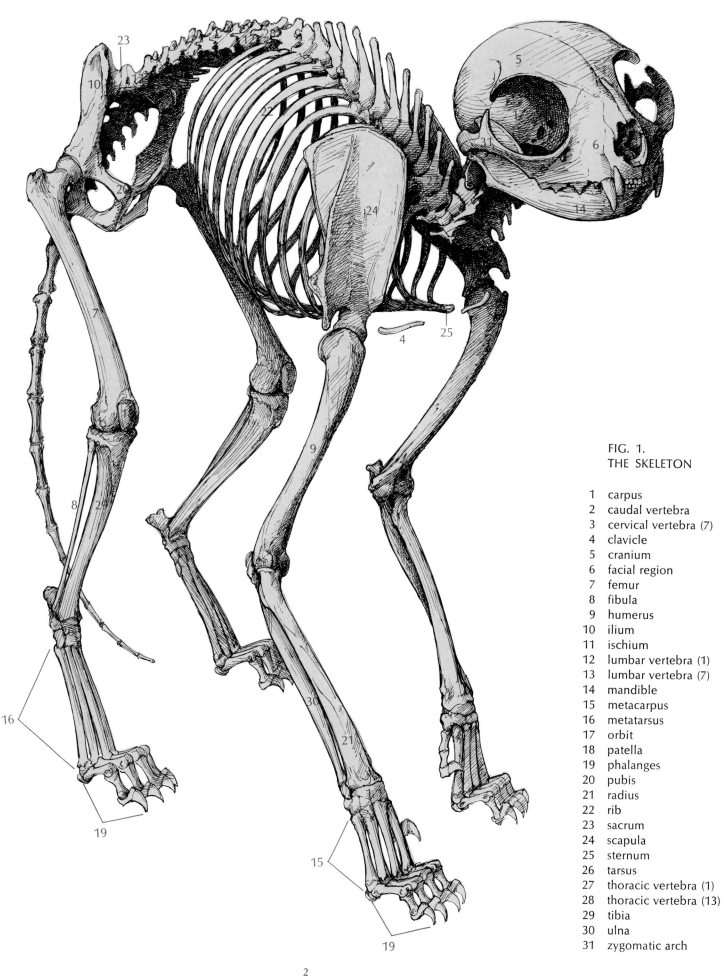

FIG. 1.
THE SKELETON

1   carpus
2   caudal vertebra
3   cervical vertebra (7)
4   clavicle
5   cranium
6   facial region
7   femur
8   fibula
9   humerus
10 ilium
11 ischium
12 lumbar vertebra (1)
13 lumbar vertebra (7)
14 mandible
15 metacarpus
16 metatarsus
17 orbit
18 patella
19 phalanges
20 pubis
21 radius
22 rib
23 sacrum
24 scapula
25 sternum
26 tarsus
27 thoracic vertebra (1)
28 thoracic vertebra (13)
29 tibia
30 ulna
31 zygomatic arch

THE SKELETON

STRUCTURE OF BONE Examine a longitudinal section of the femur and observe
that the shaft consists of a hollow tube of compact osseous tissue,
and that the proximal and distal ends of the bone are filled
by a network of slender bony spicules. Such networks are
termed cancellous bone. Most bones consist of compact and
cancellous bone, combined to provide optimum strength with
a minimum of material.

A fibrous membrane, the periosteum, covers the surface of
each bone, being absent only at the cartilaginous articulating
surfaces. The interstices of the cancellous bone and the medullary
cavities of the long bones contain marrow. Each bone is supplied
by arteries, veins, and lymphatics, which pass through nutrient
foramina to reach the marrow.

Bones are united by the following types of joints:

Immovable (synarthroses). Example: the sutures of the skull,
in which the bones are held together by interlocking margins
united by fibrous tissue.

Movable (diarthroses). Example: the knee, in which the
opposing ends of the bones are covered by articular cartilage and
are held together by ligaments lined by synovial membrane.
This type includes most of the joints.

Slightly movable (amphiarthroses). Example: the articulations
between the bodies of the vertebrae, in which the bones are
held together by flattened disks of fibrocartilage.

Examine a mounted skeleton and familiarize yourself with the
bones illustrated in Figure 1.

For descriptive purposes the skeleton may be divided into
the axial skeleton, consisting of the skull, vertebral column,
ribs, and sternum; and the appendicular skeleton, consisting of
the pectoral girdle, pelvic girdle, and limb bones.

Examine a skull and refer to Figures 2 through 8 as you
read the following description.

SKULL The skull is composed of two regions: the cranium, consisting
of the bones that enclose the brain and ear, and the facial
region, consisting of the bones that support the nose, eyes, and
mouth. Look into the external nares and observe the intricately
folded nasal conchae, which support the olfactory epithelium.

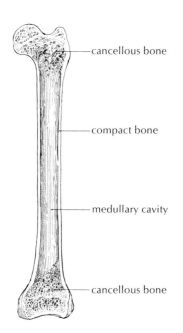

cancellous bone

compact bone

medullary cavity

cancellous bone

longitudinal section of femur

FIG. 2.
THE SKULL

1   angular process of dentary bone
2   basisphenoid bone
3   body of dentary bone
4   canine tooth
5   condyloid process of dentary bone
6   external nares
7   frontal bone
8   incisor teeth
9   interparietal bone
10   lacrimal bone
11   malar bone
12   maxilla
13   mental foramen
14   molar tooth
15   nasal bone
16   occipital bone
17   orbit
18   palatine bone
19   parietal bone
20   premaxilla
21   premolar teeth
22   presphenoid bone
23   temporal bone
24   vomer

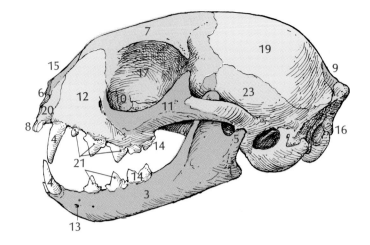

lateral view

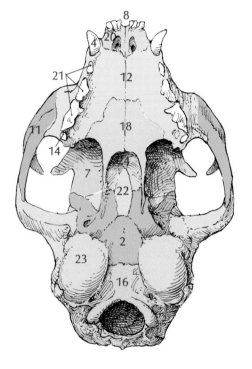

ventral view

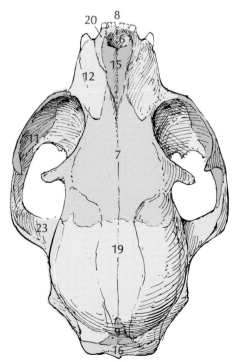

dorsal view

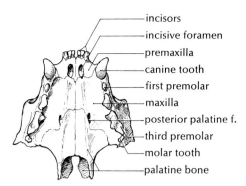

incisors
incisive foramen
premaxilla
canine tooth
first premolar
maxilla
posterior palatine f.
third premolar
molar tooth
palatine bone

hard palate, ventral view

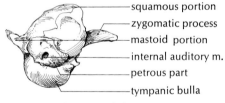

squamous portion
mastoid portion
zygomatic process
stylomastoid f.
mastoid process
external auditory m.
mandibular fossa

left temporal bone, lateral view

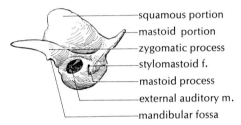

squamous portion
zygomatic process
mastoid portion
internal auditory m.
petrous part
tympanic bulla

left temporal bone, medial view

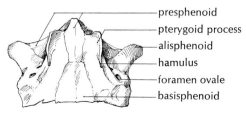

presphenoid
pterygoid process
alisphenoid
hamulus
foramen ovale
basisphenoid

presphenoid and basisphenoid
bones, ventral view

The bony socket enclosing the eye is termed the orbit. Below the orbit is the zygomatic arch composed of the malar bone, the zygomatic process of the maxilla, and the zygomatic process of the temporal bone. On the inferior surface of the zygomatic process of the temporal bone find the mandibular fossa, a depression for the articulation of the condyloid process of the dentary bone. The two dentary bones are fused in front by the mandibular symphysis: together they constitute the mandible. Posterior to the orbit, and separated from it by the postorbital processes, is the temporal fossa, which is filled in life by the temporal muscle.

Observe the hard palate, a bony partition that separates the nasal cavity from the mouth. It is formed by portions of the palatine, maxilla, and premaxilla. At the posterior end of the hard palate are the internal nares, through which air passes from the nasal cavities to the pharynx.

Look into the foramen magnum and observe the cribriform plate of the ethmoid bone at the anterior end of the cranial cavity. Olfactory nerves pass from the olfactory bulbs through the numerous small holes in the cribriform plate to reach the olfactory epithelium of the nasal cavity.

Look into the external auditory meatus to see the cavity of the middle ear. In life the external auditory meatus is closed by the tympanic membrane. Pass a small wire into the canal for the Eustachian tube to establish the connection between this canal and the middle ear cavity. In most specimens the delicate auditory ossicles may be seen within the middle ear cavity, and the round window will be seen at the posterior end of the cavity (see Fig. 75 on p. 107).

Trace the sutures of the skull and establish the contour of each bone. Also study a disarticulated skull to form an understanding of the shapes of the bones and their relationships.

The temporal bone represents the fusion of several elements that are separate in lower vertebrates. The squamous part of the temporal bone forms the zygomatic process and part of the lateral wall of the cranium. The tympanic part forms the tympanic bulla and the walls of the cavity of the middle ear. The petromastoid portion of the temporal bone consists of two parts: the mastoid process, visible in the lateral view of the skull, and the petrous part, visible in the sagittal section. The petrous part contains the inner ear.

The interparietal bone fuses with the occipitals and parietals in mature individuals and can be observed best in the skull of a young cat.

The sphenoid bone consists of three parts: a central basisphenoid and two lateral alisphenoids. The pterygoid processes, which lie on either side of the presphenoid bone, are also parts of the basisphenoid. The presphenoid, a separate bone which may be seen in the inferior view and in the lateral view, forms part of the medial wall of the orbit.

Look into the internal nares and observe that the vomer extends dorsally in the midline to form part of the partition between the two sides of the nasal cavity.

Examine the two halves of a sagittally sectioned skull to see the perpendicular plate of the ethmoid, a thin bony partition

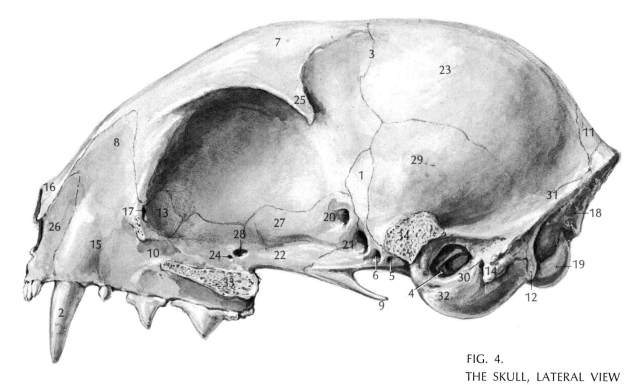

FIG. 4.
THE SKULL, LATERAL VIEW

1 alisphenoid
2 canine tooth
3 coronal suture
4 external auditory meatus
5 foramen ovale
6 foramen rotundum
7 frontal bone
8 frontal process of maxilla
9 hamulus of
    pterygoid process
10 infraorbital foramen
11 interparietal bone
12 jugular process of
    occipital bone
13 lacrimal bone
14 mastoid process
15 maxilla
16 nasal bone
17 nasolacrimal canal
18 occipital bone
19 occipital condyle
20 optic foramen
21 orbital fissure
22 palatine bone
23 parietal bone
24 posterior palatine canal
25 postorbital process
26 premaxilla
27 presphenoid bone
28 sphenopalatine foramen
29 squamous portion of
    temporal bone
30 stylomastoid foramen
31 superior nuchal line
32 tympanic bulla
33 zygomatic process of
    maxilla
34 zygomatic process of
    temporal bone

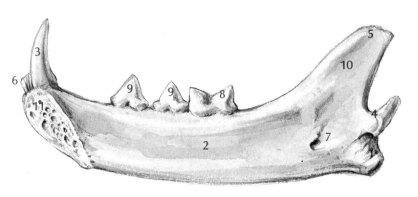

FIG. 3.
THE DENTARY BONE, MEDIAL VIEW

1 angular process
2 body
3 canine tooth
4 condyloid process
5 coronoid process
6 incisors
7 mandibular foramen
8 molar tooth
9 premolar teeth
10 ramus
11 symphysis

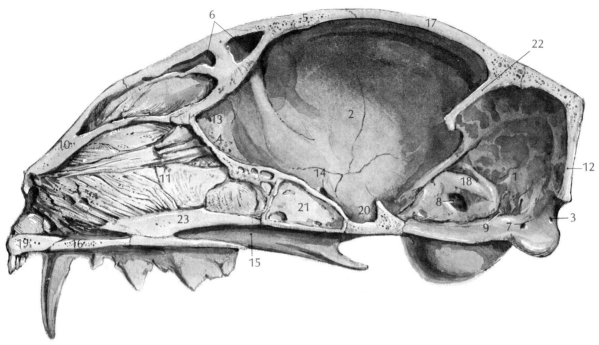

FIG. 5.

SAGITTAL SECTION OF THE SKULL

1	cerebellar fossa	10	nasal bone	18	petrous part of petromastoid
2	cerebral fossa	11	nasal conchae	19	premaxilla
3	condyloid canal	12	occipital bone	20	sella turcica of basisphenoid
4	cribriform plate of ethmoid	13	olfactory fossa	21	sphenoidal sinus of presphenoid
5	frontal bone	14	optic foramen	22	tentorium
6	frontal sinuses	15	palatine bone	23	vomer
7	hypoglossal foramen	16	palatine process of maxilla		
8	internal auditory meatus	17	parietal bone		
9	jugular foramen				

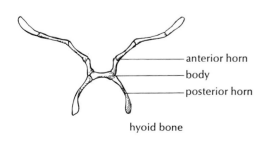

hyoid bone

that lies in the median sagittal plane and divides the nasal cavity into two equal parts. In life this bony partition is continued anteriorly by a cartilaginous partition.

The ethmoid bone consists of the cribriform plate, the perpendicular plate, and the posterior nasal conchae. Other areas of the nasal conchae constitute portions of the maxilla and the nasal bone.

After memorizing the names of the bones, study the foramina and other structures illustrated in Figures 4, 5, 6, and 7. Examine the sagittal section of the skull and determine the internal and external opening of each foramen by passing a thin wire through it. As you do this you will find it helpful to review the cranial nerves and to associate them with the foramina through which they pass (see Figure 8 and Figure 64, p. 87).

The hyoid bone is a delicate structure which lies at the anterior end of the larynx and gives attachment to certain muscles of the tongue and larynx. It consists of a short body, to which are attached anterior and posterior horns. The anterior horn has four articulated sections; the terminal section is attached to the tympanic bulla. The posterior horn is a single section attached to the larynx.

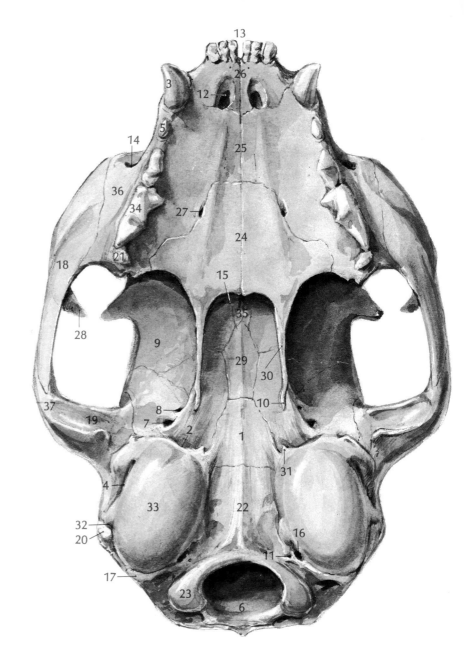

FIG. 6.

THE SKULL, VENTRAL VIEW

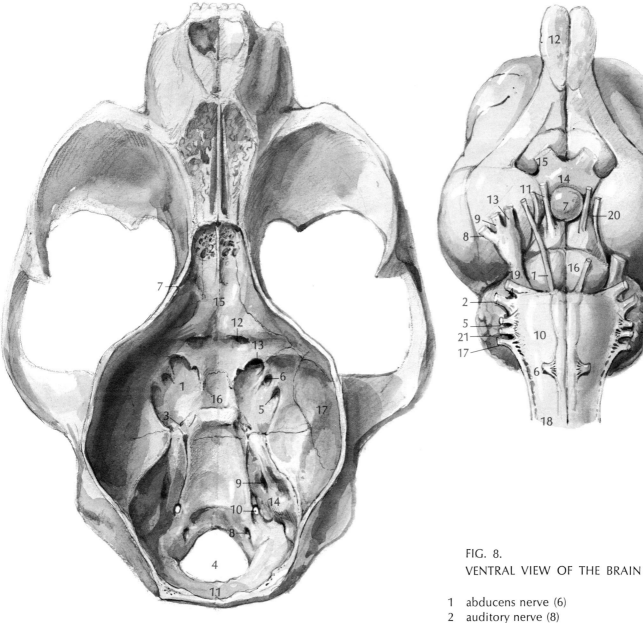

FIG. 7.

THE FLOOR OF THE CRANIAL CAVITY

1   basisphenoid
2   cribriform plate of ethmoid
3   foramen lacerum
4   foramen magnum
5   foramen ovale
6   foramen rotundum
7   frontal bone
8   hypoglossal foramen
9   internal auditory meatus
10 jugular foramen
11 occipital bone
12 optic foramen
13 orbital fissure
14 petrous part of
    petromastoid
15 presphenoid
16 sella turcica
17 temporal bone

FIG. 8.

VENTRAL VIEW OF THE BRAIN

1   abducens nerve (6)
2   auditory nerve (8)
3   cerebellum
4   facial nerve (7)
5   glossopharyngeal nerve (9)
6   hypoglossal nerve (12)
7   hypophysis
8   mandibular branch of
    trigeminal
9   maxillary branch of
    trigeminal
10 medulla oblongata
11 oculomotor nerve (3)
12 olfactory nerve (1)
13 ophthalmic branch of
    trigeminal
14 optic chiasm
15 optic nerve
16 pons
17 spinal accessory nerve (11)
18 spinal cord
19 trigeminal nerve (5)
20 trochlear nerve (4)
21 vagus nerve (10)

The vertebral column consists of five regions: cervical (seven vertebrae), thoracic (thirteen vertebrae), lumbar (seven vertebrae), sacral (three vertebrae), and caudal (about twenty vertebrae). In life the vertebrae are connected by flattened disks of fibrocartilage, strong ligaments, and complex muscles.

Examine a set of disarticulated vertebrae and learn to identify a vertebra selected at random as cervical, thoracic, lumbar, sacral, or caudal. Referring to a mounted skeleton, arrange the vertebrae in order and identify the parts labeled in Figure 9. Fit the skull against the atlas to see the articulation of the occipital condyles with the anterior articular processes (prezygapophyses) of the atlas. The atlas is atypical in that it has no body and no spinous process. Each wing of the atlas is perforated by a vertebrarterial canal through which the vertebral artery passes. This canal is found in the first six cervical vertebrae. In the atlas there is also a small atlantal foramen just above the anterior articular process for the passage of the first cervical nerve.

The odontoid process (peculiar to the axis) represents the body of the atlas, which fuses with the axis.

Put several cervical vertebrae together and observe the manner in which the articular facets fit each other. Compare the relative size of the body and the vertebral arch in a typical cervical vertebra with the bodies and the vertebral arches of typical

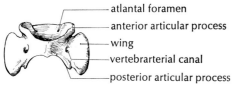

atlas, ventral view

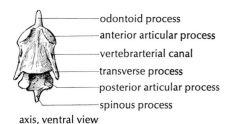

axis, ventral view

FIG. 9.

TYPICAL VERTEBRAE AND RIB

1 angle
2 anterior articular process (prezygapophysis)
3 articular surface
4 body
5 costal groove
6 dorsal arch
7 head
8 neck
9 odontoid process
10 posterior articular process (postzygapophysis)
11 sacral canal
12 shaft
13 spinous process
14 sternal end
15 transverse process
16 tubercle
17 vertebral foramen
18 wing

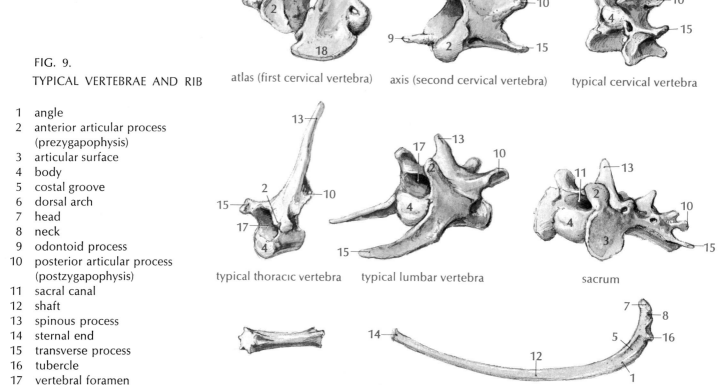

atlas (first cervical vertebra)   axis (second cervical vertebra)   typical cervical vertebra

typical thoracic vertebra   typical lumbar vertebra   sacrum

typical caudal vertebra   typical rib

10

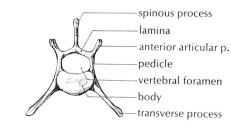

lumbar vertebra, anterior view

- spinous process
- lamina
- anterior articular p.
- pedicle
- vertebral foramen
- body
- transverse process

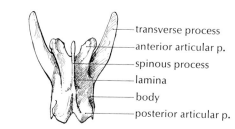

lumbar vertebra, dorsal view

- transverse process
- anterior articular p.
- spinous process
- lamina
- body
- posterior articular p.

APPENDICULAR SKELETON

FEMUR

thoracic and lumbar vertebrae. Also compare the shapes and positions of the articular facets in the cervical, thoracic, and lumbar vertebrae. Interpret the observed differences in terms of the structural requirements and the kinds of movement permitted in the various parts of the vertebral column.

Comparing the disarticulated vertebrae with the articulated skeleton, fit two thoracic vertebrae together with a rib. Observe that the head of the rib articulates with facets on both vertebral bodies and that the tubercle of the rib fits into an articular facet on the transverse process of the more caudal of the two vertebrae. The last three thoracic vertebrae have but one articular facet, located toward the anterior part of the body. Ventrally, ribs 1-9 are attached to the sternum by costal cartilages. Ribs 10, 11, and 12 are attached to the ninth costal cartilage, and rib 13 is unattached.

Examine the sternum in the articulated skeleton. It consists of eight separate units termed sternebrae. The first sternebra is the manubrium; the six central sternebrae constitute the body of the sternum; the terminal sternebra is the xiphisternum.

Place several lumbar vertebrae together and observe the form of the articular facets. Between each pair of vertebral bodies observe the paired openings, intervertebral foramina, through which the spinal nerves exit from the vertebral canal. Intervertebral foramina are found between every pair of vertebrae from the first and second cervical through the eighth and ninth caudal.

The sacrum consists of three vertebrae fused together. Observe the similarity between the last sacral vertebra and the first caudal vertebra. Note that the neural arches, transverse processes, and articular processes undergo progressive degeneration in the tail.

Study the bones of the forelimbs and hindlimbs as seen in the articulated skeleton. Select the scapulae, innominate bones, and long bones of the limbs from your set of disarticulated bones. Comparing the separate bones with the articulated skeleton, learn to distinguish the bones of the right side from those of the left. Fit the disarticulated bones together and observe the forms of the articular surfaces. You should be able to identify the articular surfaces of each bone and to match these surfaces with the corresponding articular surfaces of the bones with which it is associated.

Examine the femur. On the posterior side near the proximal and distal articular surfaces find the nutrient foramina, which admit vessels and nerves to the medullary cavity. Similar foramina will be found in many other bones. If your skeleton was made from a relatively young animal, you will see epiphyseal lines between the ends of the femur and the shaft. Similar lines may be seen in other long bones. The epiphyseal lines represent the junction between the centers of ossification in the ends (epiphyses) and the shaft (diaphysis) of the bone.

Identify and memorize the parts of the bones by referring to Figures 10, 11, 12, and 13.

The difficulty of identifying the disarticulated bones of the tarsus and carpus is too great to justify spending the required time, and you should therefore identify the bones of the feet as articulated units.

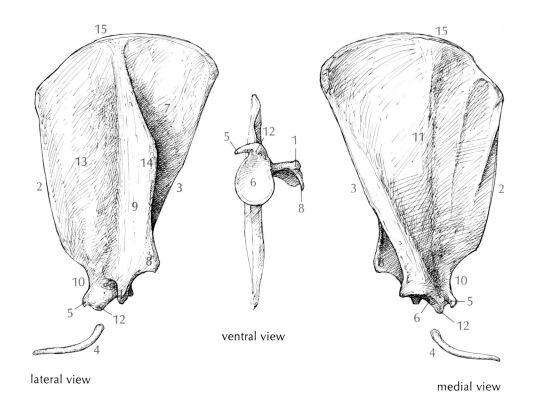

ventral view

lateral view

medial view

FIG. 10A.
THE LEFT SCAPULA
AND CLAVICLE

1 acromion
2 anterior border
3 axillary border
4 clavicle
5 coracoid process
6 glenoid fossa
7 infraspinous fossa
8 metacromion
9 spine
10 scapular notch
11 subscapular fossa
12 supraglenoid tubercle
13 supraspinous fossa
14 tuberosity of spine
15 vertebral border

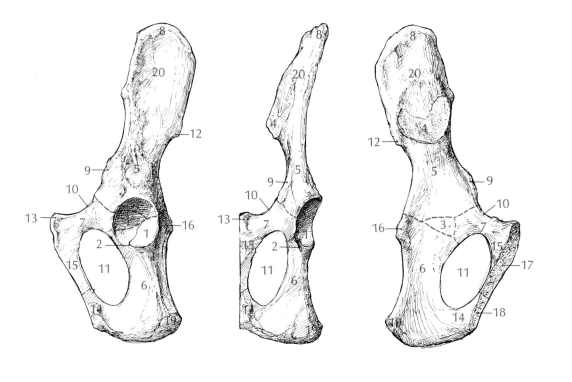

lateral view

anterior view

medial view

FIG. 10B.
THE LEFT
INNOMINATE BONE

1 acetabulum
2 acetabular notch
3 acetabular bone
4 auricular impression
5 body of ilium
6 body of ischium
7 body of pubis
8 crest of ilium
9 iliopectineal eminence
10 iliopectineal line
11 obturator foramen
12 post. inferior spine
13 pubic tubercle
14 ramus of ischium
15 ramus of pubis
16 spine of ischium
17 symphysis pubis
18 symphysis ischii
19 tuberosity of ischium
20 wing of ilium

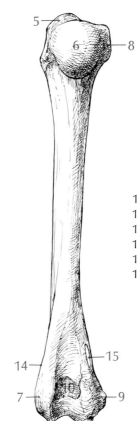

FIG. 11A.
THE LEFT HUMERUS

1 bicipital groove
2 capitulum
3 coronoid fossa
4 deltoid ridge
5 greater tuberosity
6 head
7 lateral epicondyle
8 lesser tuberosity
9 medial epicondyle
10 olecranon fossa
11 pectoral ridge
12 radial fossa
13 trochlea
14 supracondyloid ridge
15 supracondyloid foramen

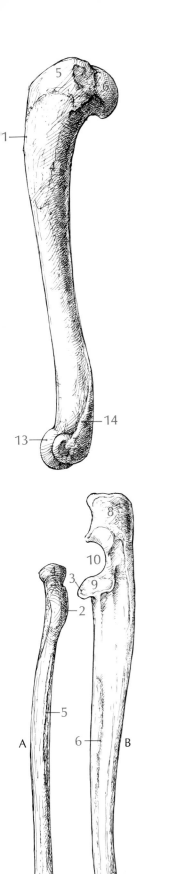

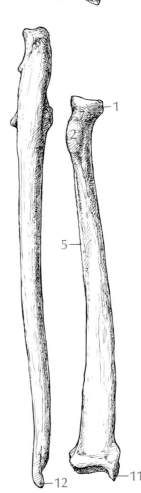

FIG. 11B.
THE LEFT RADIUS
AND ULNA

1 articular circumference
2 bicipital tuberosity
3 coronoid process
4 head
5 interosseous crest of radius
6 interosseous crest of ulna
7 neck
8 olecranon
9 radial notch
10 semilunar notch
11 styloid process of radius
12 styloid process of ulna
A radius
B ulna

lateral view   anterior view   posterior view

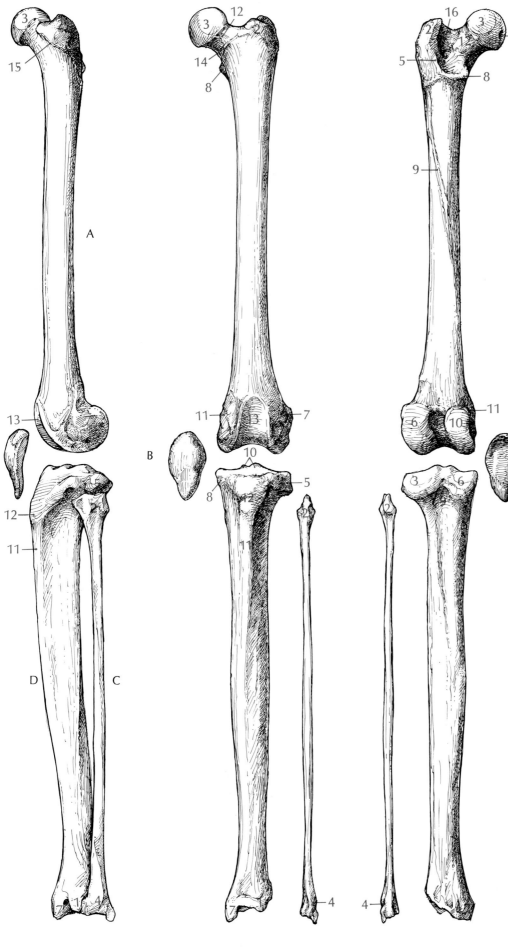

FIG. 12A.
THE LEFT FEMUR
AND PATELLA

1 fovea capitis
2 greater trochanter
3 head
4 intercondyloid fossa
5 intertrochanteric line
6 lateral condyle
7 lateral epicondyle
8 lesser trochanter
9 linea aspera
10 medial condyle
11 medial epicondyle
12 neck
13 patellar surface
14 spiral ridge
15 transverse line
16 trochanteric fossa
A femur
B patella

FIG. 12B.
THE LEFT TIBIA
AND FIBULA

1 dorsal projection
2 head
3 lateral condyle
4 lateral malleolus
5 lateral tuberosity
6 medial condyle
7 medial malleolus
8 medial tuberosity
9 popliteal notch
10 spine
11 tibial crest
12 tibial tuberosity
C fibula
D tibia

lateral view   anterior view   posterior view

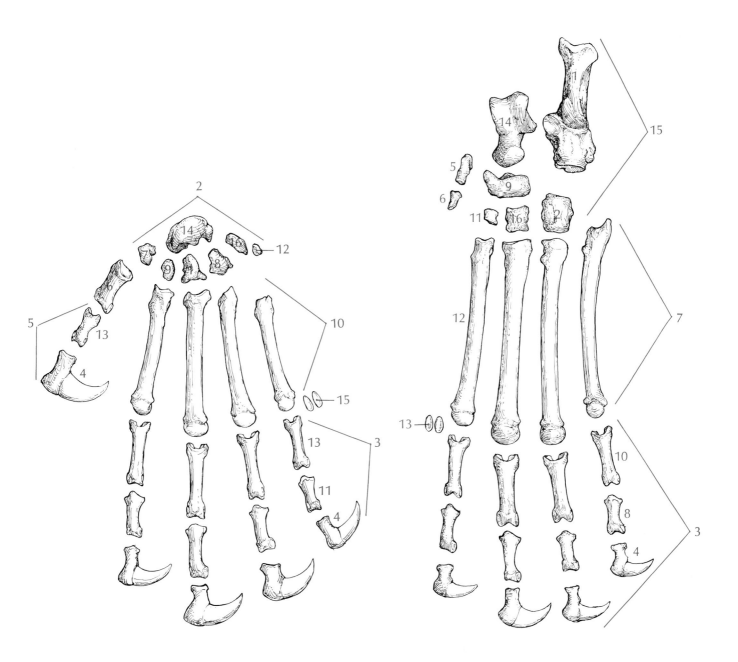

FIG. 13A.
BONES OF THE LEFT
FOREFOOT, DORSAL VIEW

1 capitate
2 carpals
3 digits
4 distal phalanx
5 first digit
6 first metacarpal
7 greater multangular
8 hamate
9 lesser multangular
10 metacarpals
11 middle phalanx
12 pisiform
13 proximal phalanx
14 scapholunar
15 sesamoid
16 triquetral

FIG. 13B.
BONES OF THE LEFT
HINDFOOT, DORSAL VIEW

1 calcaneus
2 cuboid
3 digits
4 distal phalanx
5 first cuneiform
6 first metatarsal
7 metatarsals
8 middle phalanx
9 navicular
10 proximal phalanx
11 second cuneiform
12 second metatarsal
13 sesamoid
14 talus
15 tarsals
16 third cuneiform

THE MUSCLES

STRUCTURE OF MUSCLE   The fleshy central portion of a skeletal or voluntary muscle consists of contractile muscle fibers and is termed the belly. It is usually attached at both ends by connective tissue fibers which form either a tendon or a flat sheet termed an aponeurosis. In some cases muscle fibers attach directly to the periosteum without the intervention of a tendon. The more stable of the two attachments is termed the origin; the more mobile attachment is termed the insertion. The origin is usually nearer the sagittal plane than the insertion. In some cases (the intercostal muscles, for instance) both attachments are equally mobile and neither is nearer the sagittal plane. In such instances the terms *origin* and *insertion* may be assigned arbitrarily even though they do not have their usual functional significance.

ATTACHMENT   Most muscles are attached to bones, cartilages, or ligaments. Muscles may also attach to the fascia covering another muscle, to the mucous membrane (tongue muscles), to the skin (facial muscles), or they may form circular bands (sphincters).

ACTIONS   Muscles act only by contraction and almost always occur in antagonistic groups. Examples of such groups are: extensors, which straighten joints, and flexors, which bend them; adductors, which move appendages toward the median sagittal plane, and abductors, which move them away; pronators, which turn the dorsal surface of a limb anteriorly, and supinators, which turn the ventral surface of a limb anteriorly; levators, which raise structures, and depressors, which lower them; sphincters, which close openings, and dilators, which open them.

For descriptive purposes we usually speak of a muscle as having a primary action, determined by the origin and insertion, and a number of secondary actions, determined by the activities of other muscles. Only the primary action of the muscle is listed in the descriptions which follow, and in many cases the accounts of the origins and insertions are simplified. Most of the muscles you will study in the cat have homologs of the same name in the human. When this is not the case, or when the human homolog is significantly different, it will be noted in the description of the muscle. Refer to the skeleton of the cat and to drawings or models of the human muscles during your dissection.

DISSECTING TECHNIQUE   Before beginning the dissection of a given area, read through the dissecting instructions and study the illustrations in order to familiarize yourself with the structures you will encounter. Then study and identify the individual muscles as they are listed in the text, freeing each one from the surrounding muscles and confirming its origin and insertion as you do so. In order to define the limits of a muscle, use small scissors and forceps to trim away the overlying fat and fascia until you can see the direction of the muscle fibers. Look for a change in the direction of the fibers near the place where the edge of the muscle should be, and attempt to slip the flat edge of your scalpel handle between two separate layers of muscle at this point. If one layer separates easily from another you have your scalpel between two different muscles. Do not try to cut or force the separation of the muscles. If you are looking in the right place the separation will be natural. When it is necessary to cut a muscle in order to expose underlying structures, make your cut at right angles to the direction of the fibers and about half way between the origin and the insertion. Then pull back the two halves of the muscle and dissect them away from the underlying structures. After verifying the points of attachment you may wish to remove the muscle completely in order to make a neat dissection. In some cases it is desirable to leave the cut portions of the superficial muscles intact so that they can be replaced to re-establish their natural relationships with the underlying structures. It is often helpful to confirm an observation made on the left side of the specimen by dissecting on the right. If you do this, leave intact the origins and insertions of any muscles you cut on the right and do not destroy any of the nerves or vessels, which will be studied later.

SKINNING   Place the cat on its back. Make a midventral cut through the skin and extend it from the jaw to the anus. If your cat is a female, cut around the nipples and dissect the skin away from the mammary glands, which lie between the skin and the superficial muscles of the abdomen.

Separate the skin from the underlying muscles by blunt dissection. Make additional cuts as necessary to remove the skin from the torso and legs, leaving it intact around the mouth, eyes, ears, and feet.

FASCIA   As you pull the skin away from the body you will observe that it is connected to the underlying structures by a white, fibrous membrane consisting of elastic fibers and fat. This layer of connective tissue, termed the superficial fascia, is distinguished from the deep fascia, the tough fibrous membrane which invests the individual muscles.

CUTANEOUS MAXIMUS   In the lateral thoracic area you will observe muscle fibers lying within the superficial fascia and attached directly to the skin. These fibers are part of the cutaneous maximus, which originates from the muscles of the axilla and from the ventral side of the abdomen and thorax, and inserts on the skin. It serves to twitch the skin, as in avoiding irritants. Cut through the cutaneous maximus

PLATYSMA   and remove it with the skin. Another muscle which moves the skin will be found in the area of the neck and face. This is the platysma, which inserts on the skin around the ears, eyes, and lips. Remove it with the skin. In skinning the head and neck, refer

17

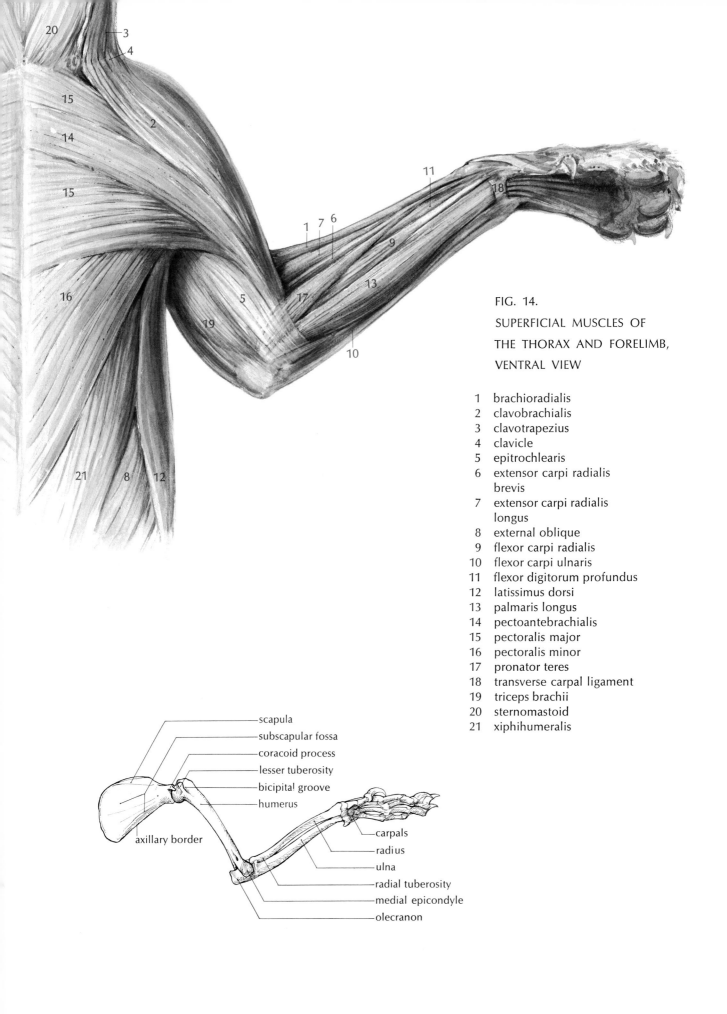

FIG. 14.

SUPERFICIAL MUSCLES OF
THE THORAX AND FORELIMB,
VENTRAL VIEW

1   brachioradialis
2   clavobrachialis
3   clavotrapezius
4   clavicle
5   epitrochlearis
6   extensor carpi radialis
    brevis
7   extensor carpi radialis
    longus
8   external oblique
9   flexor carpi radialis
10 flexor carpi ulnaris
11 flexor digitorum profundus
12 latissimus dorsi
13 palmaris longus
14 pectoantebrachialis
15 pectoralis major
16 pectoralis minor
17 pronator teres
18 transverse carpal ligament
19 triceps brachii
20 sternomastoid
21 xiphihumeralis

scapula
subscapular fossa
coracoid process
lesser tuberosity
bicipital groove
humerus
carpals
radius
ulna
radial tuberosity
medial epicondyle
olecranon
axillary border

to Figures 15 and 16 and be careful not to damage the superficial veins and glands. As you skin the dorsal side of the thorax and abdomen observe the segmentally arranged cutaneous vessels and nerves passing from the dorsal body wall to the skin.

Examine the ventral aspect of the thorax and forelimb, and identify the muscles illustrated in Figure 14.

The pectoantebrachialis is the most superficial of the chest muscles. Separate it from the underlying pectoralis major, observing that it originates from the manubrium and inserts on the fascia of the forelimb near the elbow. It has no homolog in man. Cut the pectoantebrachialis in the middle and pull back both ends. Dissect the clavobrachialis and the clavotrapezius away from the humerus to expose the insertion of the pectoralis major.

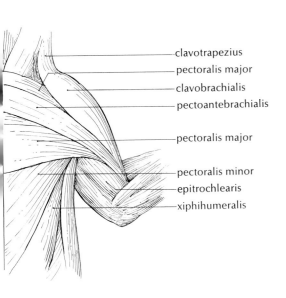

- clavotrapezius
- pectoralis major
- clavobrachialis
- pectoantebrachialis

- pectoralis major

- pectoralis minor
- epitrochlearis
- xiphihumeralis

The pectoralis major and minor are somewhat variable in form. If the pectoralis major is not readily distinguishable from the pectoralis minor, examine their insertions on the humerus. The fibers of the pectoralis major pass from the sternum, almost at right angles to the midline of the body, and insert on the proximal two thirds of the humerus between the biceps and the brachialis. The pectoralis minor crosses obliquely, deep to the pectoralis major, and inserts on the proximal half of the humerus. In the human the pectoralis major is much larger than the pectoralis minor. This is not the case in the cat.

The xiphihumeralis originates from the sternum, passes obliquely deep to the pectoralis minor, and inserts by a narrow tendon near the proximal end of the humerus. The xiphihumeralis has no homolog in man. The pectoantebrachialis, pectoralis major, pectoralis minor, and xiphihumeralis act together to rotate the forelimb and to draw the forelimb toward the chest.

The epitrochlearis is a thin, superficial muscle which appears to be an extension of the latissimus dorsi. It originates from the lateral surface of the ventral border of the latissimus dorsi and inserts by a thin aponeurosis which is continuous with the fascia of the lower forelimb. It acts in common with the triceps as an extensor of the elbow. It has no homolog in man.

The pronator teres originates from the medial epicondyle of the humerus and inserts about the middle of the radius. It rotates the radius to the prone position.

The palmaris longus arises from the medial epicondyle of the humerus and passes under the transverse carpal ligament. Trace its tendons, which insert on the pads of the forefoot and on the proximal phalanges of the digits. It is a flexor of the wrist and digits. The palmaris longus of the human is a slender muscle which inserts on the fascia of the palm. It is absent in about 10 per cent of humans.

The flexor carpi radialis originates from the medial epicondyle of the humerus and inserts on the second and third metacarpals. It flexes the wrist.

The flexor carpi ulnaris arises as two heads. One originates from the medial epicondyle of the humerus; the other originates from the olecranon. About the middle of the ulna the two heads join and pass along the ulnar border of the lower forelimb to insert on the ulnar side of the carpals. This muscle acts as a flexor of the wrist.

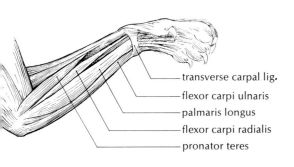

- transverse carpal lig.
- flexor carpi ulnaris
- palmaris longus
- flexor carpi radialis
- pronator teres

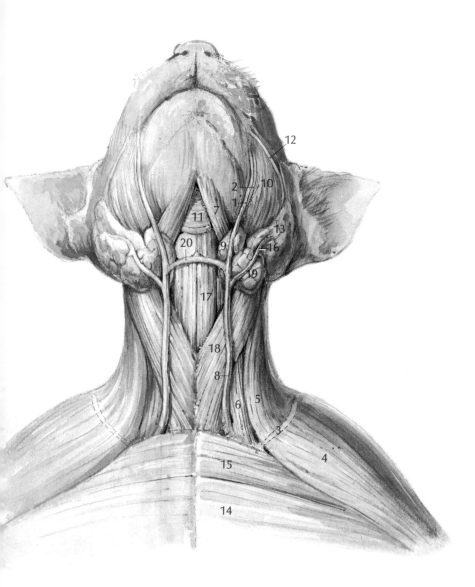

FIG. 15.
SUPERFICIAL MUSCLES OF
THE NECK, VENTRAL VIEW

1 anterior facial vein
2 branch of facial nerve
3 clavicle
4 clavobrachialis
5 clavotrapezius
6 cleidomastoid
7 digastric
8 external jugular vein
9 lymph nodes
10 masseter
11 mylohyoid
12 parotid duct
13 parotid gland
14 pectoantebrachialis
15 pectoralis major
16 posterior facial vein
17 sternohyoid
18 sternomastoid
19 submaxillary gland
20 transverse jugular vein

NECK AND SHOULDER

Examine the ventral aspect of the neck and identify the muscles illustrated in Figure 15.

The mylohyoid originates from the medial surface of the two dentary bones and inserts on a median raphe which extends from the hyoid bone to the mandibular symphysis. It raises the floor of the mouth and draws the hyoid bone anteriorly.

The digastric originates from the jugular and mastoid processes and inserts on the inferior border of the dentary bone. It is a depressor of the mandible.

The masseter originates from the zygomatic arch and inserts on the posterior half of the lateral surface of the dentary bone. It elevates the mandible.

Examine the lateral aspect of the neck and shoulder, and identify the muscles illustrated in Figure 16.

The temporal muscle originates from the lateral surface of the skull posterior to the orbit, and inserts on the coronoid process of the dentary bone. It acts with the masseter to elevate the mandible.

The sternomastoid originates from the manubrium of the sternum and from the midventral line anterior to the manubrium. It passes obliquely around the neck to insert on the superior nuchal line and on the mastoid process. Singly it turns the head; both muscles

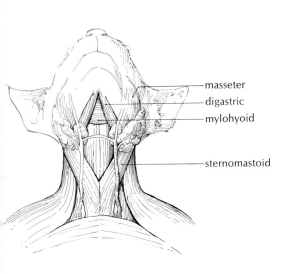

masseter
digastric
mylohyoid

sternomastoid

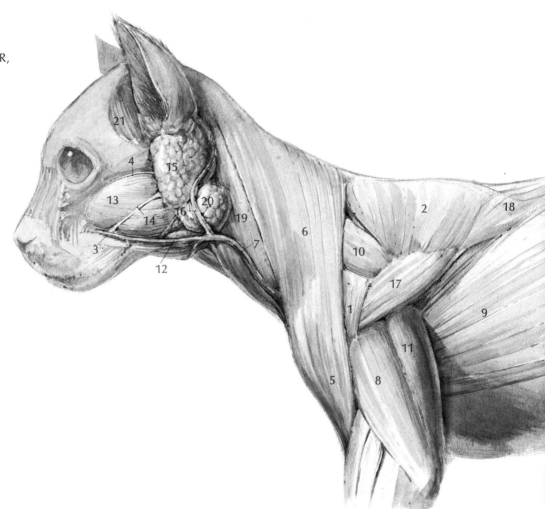

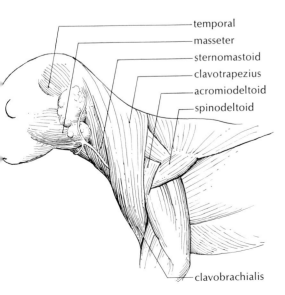

acting together depress the head and neck.

The fibers of the clavotrapezius and the clavobrachialis are continuous, but their innervation is separate. Find the position of the clavicle by palpation at the point where the fibers of the clavobrachialis meet those of the clavotrapezius. The clavotrapezius originates from the superior nuchal line of the skull and from the median dorsal line of the neck. It inserts on the clavicle. The clavobrachialis originates from the clavicle and inserts on the proximal end of the ulna. The clavotrapezius and clavobrachialis work together in the forward extension of the humerus, as in running; they also assist in turning the head and in flexing the elbow. The clavotrapezius is homologous with that portion of the trapezius which inserts on the clavicle in man; the clavobrachialis is homologous with that portion of the deltoid which originates from the clavicle in man.

The acromiodeltoid originates from the acromion. The spinodeltoid originates from the spine of the scapula. Both muscles insert on the proximal end of the humerus and work together to raise and rotate the humerus. The acromiodeltoid, spinodeltoid, and clavobrachialis are homologous with the single deltoid muscle of man.

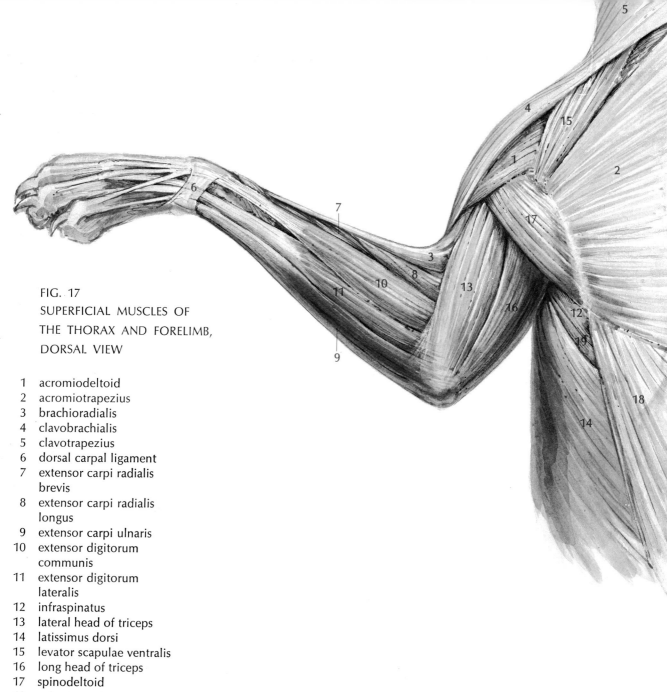

FIG. 17
SUPERFICIAL MUSCLES OF
THE THORAX AND FORELIMB,
DORSAL VIEW

1   acromiodeltoid
2   acromiotrapezius
3   brachioradialis
4   clavobrachialis
5   clavotrapezius
6   dorsal carpal ligament
7   extensor carpi radialis
    brevis
8   extensor carpi radialis
    longus
9   extensor carpi ulnaris
10 extensor digitorum
    communis
11 extensor digitorum
    lateralis
12 infraspinatus
13 lateral head of triceps
14 latissimus dorsi
15 levator scapulae ventralis
16 long head of triceps
17 spinodeltoid
18 spinotrapezius
19 teres major

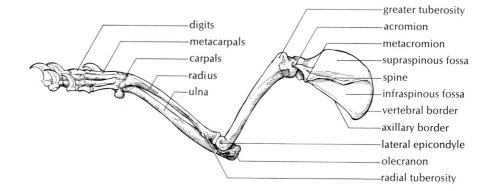

digits
metacarpals
carpals
radius
ulna

greater tuberosity
acromion
metacromion
supraspinous fossa
spine
infraspinous fossa
vertebral border
axillary border
lateral epicondyle
olecranon
radial tuberosity

Examine the dorsal aspect of the shoulder and forelimb, and identify the muscles illustrated in Figure 17.

The acromiotrapezius originates by a thin aponeurosis from the spinous processes of the cervical and anterior thoracic vertebrae. It inserts on the spine of the scapula.

The spinotrapezius originates from the spinous processes of the posterior thoracic vertebrae. It inserts on the spine of the scapula and the surrounding fascia. Together the spinotrapezius and the acromiotrapezius act to hold the two scapulae together. The spinotrapezius also draws the scapula posteriorly. The clavotrapezius, acromiotrapezius, and spinotrapezius of the cat are homologous with the single trapezius muscle of the human.

The latissimus dorsi originates from the spinous processes of the posterior thoracic vertebrae and the lumbar vertebrae. It inserts on the medial aspect of the humerus and acts to pull the forelimb dorsally and posteriorly.

Remove the clavotrapezius and the clavobrachialis to expose the levator scapulae ventralis, which originates from the occipital bone and the transverse process of the atlas. It inserts on the metacromion and nearby fascia. It acts to draw the scapula anteriorly. This muscle has no homolog in man.

The triceps brachii has three divisions. The long head originates from the axillary border of the scapula just below the glenoid fossa. The medial and lateral heads originate from the humerus. In the cat the medial head is divisible into three smaller slips which need not be individually identified. All three heads of the triceps insert on the olecranon. Cut the lateral head and pull it back to see the medial head from the dorsal view. Also see the triceps as illustrated in Figure 20, page 25, and Figure 21, page 27. The triceps acts to extend the elbow.

The brachioradialis originates about the middle of the humerus and inserts on the distal end of the radius. It rotates the radius and supinates the foot.

The extensor carpi radialis longus and the extensor carpi radialis brevis lie deep to the brachioradialis. They originate from the lateral surface of the humerus above the lateral epicondyle and insert on the bases of the second and third metacarpals, respectively. They extend the carpal joint (see Fig. 20, p. 25).

The extensor digitorum communis originates from the lateral surface of the humerus above the lateral epicondyle. It divides into four tendons which pass under the dorsal carpal ligament and insert on the bases of the second phalanges of digits 2-5. It extends the digits.

The extensor digitorum lateralis originates from the lateral surface of the humerus above the lateral epicondyle. Its three tendons insert, in common with the tendons of the extensor digitorum communis, on digits 3-5. In some specimens it will also be found to give a tendon to the second digit. It acts with the extensor digitorum communis to extend the digits. The homolog of the extensor digitorum lateralis in man is a much smaller muscle, the extensor digiti quinti proprius, which inserts only on the little finger.

The extensor carpi ulnaris originates from the lateral epicondyle of the humerus and inserts on the proximal end of the fifth metacarpal. It extends the carpal joint.

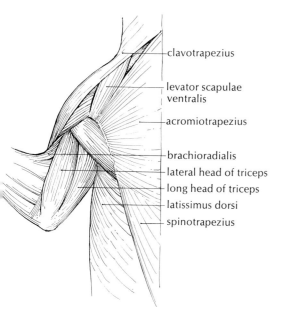

clavotrapezius

levator scapulae ventralis

acromiotrapezius

brachioradialis
lateral head of triceps
long head of triceps
latissimus dorsi
spinotrapezius

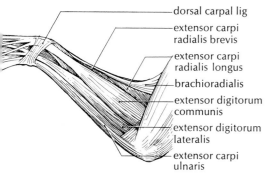

dorsal carpal lig
extensor carpi radialis brevis
extensor carpi radialis longus
brachioradialis
extensor digitorum communis
extensor digitorum lateralis
extensor carpi ulnaris

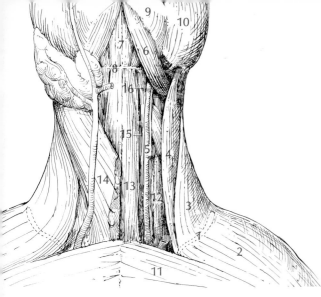

Refer to Figure 18. Remove the external jugular vein and the salivary glands and lymph nodes from the left side of the neck. Remove the sternomastoid and mylohyoid muscles, and identify the cleidomastoid, which lies deep to the sternomastoid and the clavotrapezius.

The cleidomastoid originates from the clavicle and inserts on the mastoid process. It acts together with the sternomastoid and the clavotrapezius in effecting movements of the head and forelimb. In man the homologs of the sternomastoid and the cleidomastoid are united to form a single muscle, the sternocleidomastoid.

The sternohyoid is a slender muscle in the midline of the throat. It originates from the first costal cartilage, inserts on the hyoid bone, and acts to draw the hyoid bone posteriorly. Determine the position of the hyoid bone by palpation.

The sternothyroid lies deep to the sternohyoid. It originates from the first costal cartilage, and inserts on the thyroid cartilage of the larynx. It draws the larynx posteriorly.

Cut the pectoralis major, the pectoralis minor, and the xiphihumeralis, examining their origins and insertions as you do so. Deep to the pectoralis minor you will find considerable fat and connective tissue surrounding the vessels and nerves which supply the forelimb. Remove the vessels, nerves, and fat to expose the muscles as illustrated in Figure 19.

The serratus ventralis originates by a series of individual slips from the first nine or ten ribs and from the transverse process of the last five cervical vertebrae. It inserts on the vertebral border of the scapula. The serratus ventralis is the largest of the muscles which attach the forelimb to the thorax. It serves to suspend the thorax, and is homologous with the serratus anterior and the levator scapulae of the human. In the cat that portion of the serratus ventralis which arises from the cervical vertebrae is sometimes termed the levator scapulae, although it is not a separate muscle as it is in man.

FIG. 18

DEEP MUSCLES OF THE NECK, VENTRAL VIEW

1 clavicle
2 clavobrachialis
3 clavotrapezius
4 cleidomastoid
5 common carotid artery and vagus nerve
6 digastric
7 geniohyoid
8 hyoid bone
9 mandible
10 masseter
11 pectoralis major
12 scalenes
13 sternohyoid
14 sternomastoid
15 sternothyroid
16 thyrohyoid

FIG. 19.

DEEP MUSCLES OF THE THORAX, VENTRAL VIEW

1 biceps brachii
2 brachialis
3 coracobrachialis
4 cut insertions of pectoralis major and pectoante-brachialis
5 epitrochlearis
6 external oblique
7 latissimus dorsi
8 levator scapulae ventralis
9 pectoralis minor
10 rectus abdominis
11 scalenus medius
12 serratus ventralis
13 sternohyoid
14 subscapularis
15 teres major
16 transversus costarum
17 triceps brachii
18 xiphihumeralis

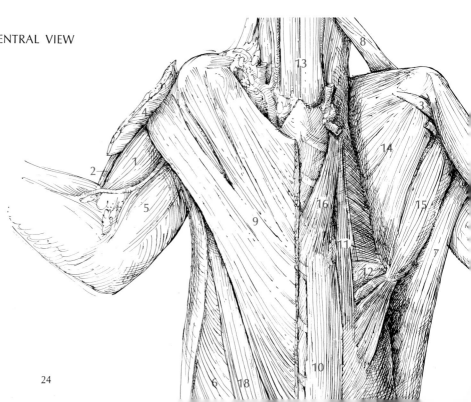

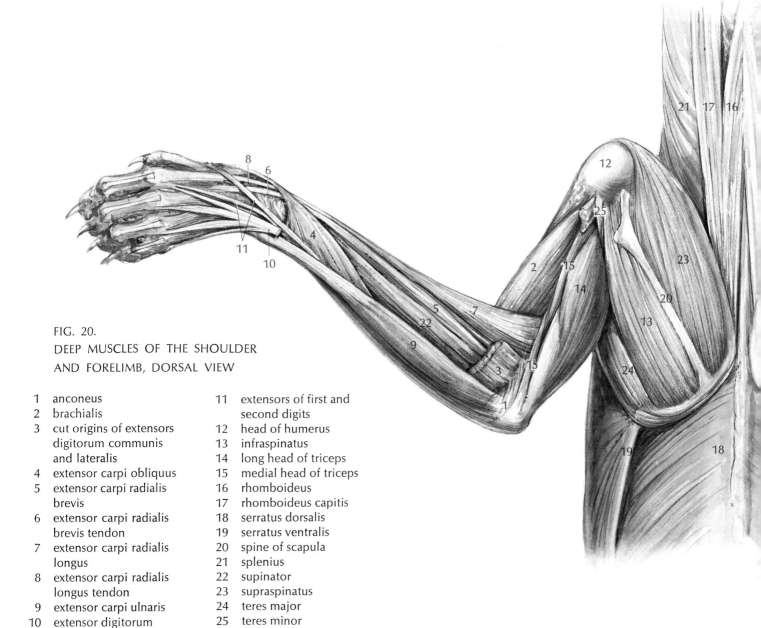

FIG. 20.

DEEP MUSCLES OF THE SHOULDER
AND FORELIMB, DORSAL VIEW

1	anconeus	11	extensors of first and second digits
2	brachialis	12	head of humerus
3	cut origins of extensors digitorum communis and lateralis	13	infraspinatus
4	extensor carpi obliquus	14	long head of triceps
5	extensor carpi radialis brevis	15	medial head of triceps
6	extensor carpi radialis brevis tendon	16	rhomboideus
7	extensor carpi radialis longus	17	rhomboideus capitis
8	extensor carpi radialis longus tendon	18	serratus dorsalis
9	extensor carpi ulnaris	19	serratus ventralis
10	extensor digitorum lateralis tendon	20	spine of scapula
		21	splenius
		22	supinator
		23	supraspinatus
		24	teres major
		25	teres minor

DISSECTION OF LATERAL FORELIMB MUSCLES

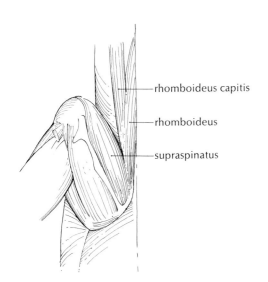

rhomboideus capitis

rhomboideus

supraspinatus

Referring to Figures 17 and 20, remove the following muscles: spinotrapezius, acromiotrapezius, levator scapulae ventralis, acromiodeltoid, spinodeltoid, lateral head of the triceps, brachioradialis, extensor digitorum communis, and extensor digitorum lateralis. Your dissection should now resemble Figure 20.

The rhomboideus originates from the spinous processes and ligaments of the posterior cervical and anterior thoracic vertebrae, and inserts on the vertebral border of the scapula. A separate portion of the rhomboideus, termed the rhomboideus capitis, originates from the superior nuchal line. The rhomboideus draws the scapula forward and toward the mid-dorsal line; the rhomboideus capitis assists in the rotation of the scapula. The rhomboideus is homologous with the rhomboideus major and minor in man, but the rhomboideus capitis has no homolog in man.

The supraspinatus originates from the supraspinous fossa of the scapula and inserts on the greater tuberosity of the humerus. It draws the humerus anteriorly.

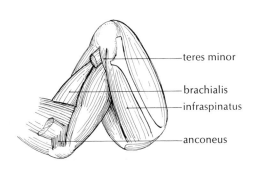

teres minor

brachialis

infraspinatus

anconeus

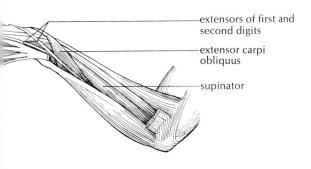

extensors of first and second digits

extensor carpi obliquus

supinator

DISSECTION OF MEDIAL FORELIMB MUSCLES

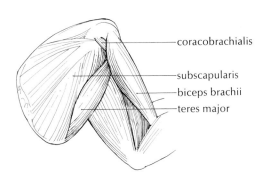

coracobrachialis

subscapularis

biceps brachii

teres major

The infraspinatus originates from the infraspinous fossa of the scapula and inserts on the lateral aspect of the greater tuberosity of the humerus. It is a lateral rotator of the humerus.

The teres minor originates from the axillary border of the scapula and inserts on the greater tuberosity of the humerus. It assists the infraspinatus.

The brachialis originates from the lateral surface of the humerus and inserts on the ulna, acting with the biceps to flex the elbow.

The anconeus consists of short superficial muscle fibers which originate near the lateral epicondyle of the humerus and insert near the olecranon. It assists in the extension of the elbow and acts on the capsule of the joint.

The extensor carpi obliquus originates from the ulna and passes around the wrist to insert on the base of the first metacarpal. It abducts the first digit and extends the wrist. It is homologous with the abductor pollicis longus of man.

The extensors of the first and second digits are small slips which originate from the lateral surface of the ulna and insert on the first and second digits. The extensor of the first digit is sometimes absent. These muscles are homologous with the extensor pollicis longus and extensor indicis proprius of man.

The supinator originates from the ligaments of the elbow and may appear continuous with the extensor carpi obliquus. It inserts on the proximal end of the radius and acts as a lateral rotator of the radius to produce supination of the forefoot.

Cut the rhomboids and the serratus ventralis at the vertebral border of the scapula, and free the forelimb from the thorax. Trim away any remaining portions of the trapezius, latissimus dorsi, and serratus ventralis to expose the underlying muscles. Referring to Figures 14 and 21, cut and remove the palmaris longus, flexor carpi radialis, and flexor carpi ulnaris. Cut the carpal ligament which holds the flexor tendons against the bones of the wrist. Your dissection should now resemble Figure 21.

The subscapularis originates from the subscapular fossa and inserts on the lesser tuberosity of the humerus. It is an adductor of the humerus.

The teres major originates from the axillary border of the scapula; it inserts in common with the latissimus dorsi on the proximal end of the humerus. It rotates the humerus and draws it posteriorly.

The coracobrachialis originates from the coracoid process and inserts on the proximal end of the humerus. It is an adductor of the humerus.

The biceps brachii originates just above the glenoid fossa of the scapula. Cut away the capsular ligament of the shoulder around the proximal end of the biceps tendon to see its origin. Observe that the tendon lies in the bicipital groove of the humerus, thus serving to stabilize the shoulder joint. The biceps inserts on the radial tuberosity near the proximal end of the radius, and acts as a flexor of the elbow. In man the biceps has two heads of origin: one above the glenoid fossa, as in the cat, and one from the coracoid process.

The flexor digitorum sublimis is a prominent muscle in man, but in the cat it is represented by a few slender strips which originate from the surface of the palmaris longus and flexor digitorum

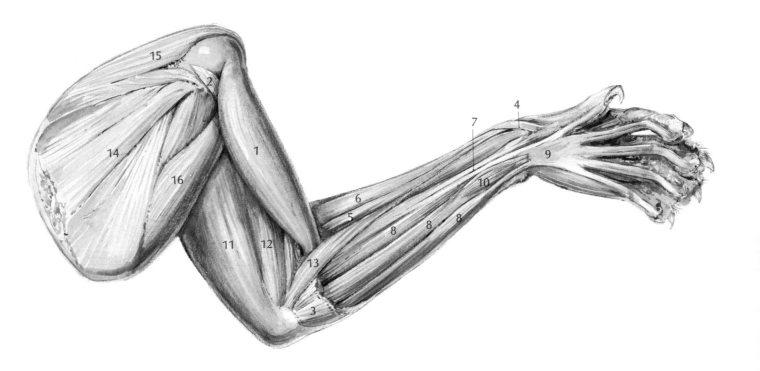

FIG. 21.

DEEP MUSCLES OF THE SHOULDER AND FORELIMB, VENTRAL VIEW

1	biceps brachii	7	flexor carpi radialis tendon
2	coracobrachialis	8	flexor digitorum profundus
3	cut origins of palmaris longus and flexor carpi ulnaris	9	flexor digitorum profundus tendon
4	extensor carpi obliquus tendon	10	flexor digitorum sublimis
5	extensor carpi radialis brevis	11	long head of triceps
		12	medial head of triceps
6	extensor carpi radialis longus	13	pronator teres
		14	subscapularis
		15	supraspinatus
		16	teres major

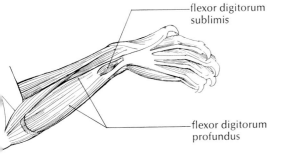

flexor digitorum sublimis

flexor digitorum profundus

profundus near the carpal joint. They insert on the digits.

The flexor digitorum profundus originates as five separate heads from the radius, the medial epicondyle of the humerus, and the ulna. It inserts by strong tendons on the distal phalanges of the digits and acts as a flexor of the digits.

The pronator quadratus (not illustrated) lies beneath the tendon of the flexor digitorum profundus, just proximal to the carpal joint. It originates from the ulna and consists of short muscle fibers which pass obliquely to the radius. It acts with the pronator teres in the medial rotation of the radius.

FIG. 22.
MUSCLES OF THE THORAX, LATERAL VIEW

1   common carotid artery
    and sympathetic trunk
2   cut insertion of serratus
    ventralis
3   external oblique
4   geniohyoid
5   masseter
6   rectus abdominis
7   rhomboideus
8   rhomboideus capitis
9   scalenus anterior
10 scalenus medius
11 scalenus posterior
12 serratus dorsalis cranialis
13 serratus ventralis
14 splenius
15 sternohyoid
16 sternothyroid
17 temporal
18 thyrohyoid
19 thyroid gland
20 transversus costarum

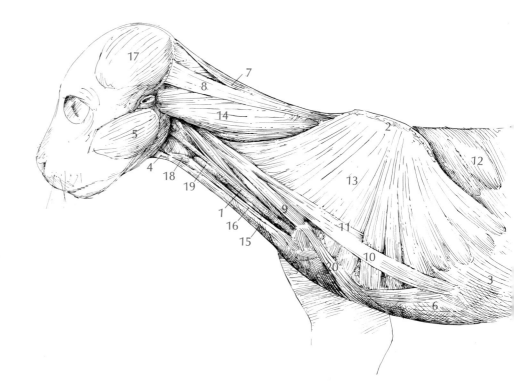

SUPERFICIAL DISSECTION OF
LATERAL THORACIC MUSCLES

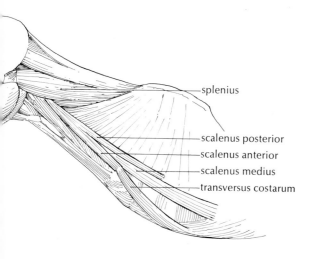

splenius

scalenus posterior
scalenus anterior
scalenus medius
transversus costarum

DEEP DISSECTION OF
LATERAL THORACIC MUSCLES

Examine the left lateral aspect of the head, neck, and thorax.
The scapula and forelimb are removed, but the rhomboids and the
serratus ventralis are still in their normal positions as seen in Figure 22.
Remove any remaining superficial fascia and fat to make a
dissection resembling the illustration.

The splenius originates from the nuchal ligament, in the mid-dorsal
line of the neck, and inserts on the superior nuchal line of the
skull. It turns or raises the head.

The scalenes are three muscles which originate on the ribs and
insert on the transverse processes of the cervical vertebrae.
The most prominent is the middle portion, the scalenus medius,
which originates from ribs 6-9. The most ventrally located is the
scalenus anterior. It originates from the second and third ribs,
and may appear to be continuous with the transversus costarum. The
scalenus posterior, which originates from the third rib, is the most
dorsally placed of the three scalenes. The scalenes bend the
neck and draw the ribs anteriorly.

The transversus costarum originates from the sternum and inserts
on the first rib and costal cartilage. It acts together with the scalenes.

Remove the rhomboids and cut the serratus ventralis near its
insertion to make a dissection similar to Figure 23.

The serratus dorsalis cranialis originates as a thin aponeurosis
overlying the anterior divisions of the sacrospinalis and inserts on ribs
1-9 by a series of thin muscular strips. The serratus dorsalis
caudalis lies posterior to the serratus dorsalis cranialis and may

FIG. 23.

DEEP MUSCLES OF THE THORAX, LATERAL VIEW

1 cut edge of serratus
  ventralis
2 external intercostal
3 external oblique
4 longus capitis
5 lumbodorsal fascia
6 scalenus anterior
7 scalenus medius
8 scalenus posterior
9 serratus dorsalis caudalis
10 serratus dorsalis cranialis
11 splenius
12 sternohyoid
13 sternothyroid
14 thyroid gland
15 transversus costarum

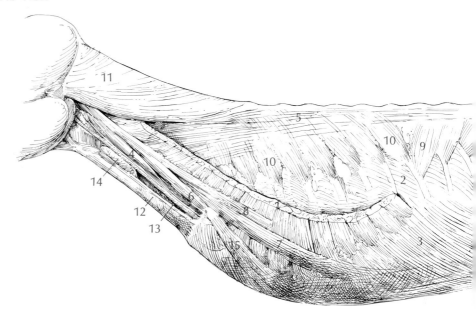

be distinguished by the anterior direction of its fibers. It consists of four or five separate slips which insert on the posterior ribs. The serratus dorsalis cranialis draws the ribs forward and outward, increasing the volume of the thorax during inspiration. The serratus dorsalis caudalis draws the ribs posteriorly, decreasing the volume of the thorax during expiration.

Remove the serratus dorsalis, scalenes, transversus costarum, splenius, and the remaining portion of the serratus ventralis.

INTERCOSTALS The external and internal intercostals extend between adjacent ribs and serve to draw the ribs together during respiration. Cut away the external intercostal muscle between two adjacent ribs. Observe that the fibers of the internal intercostals run in the opposite direction from the external intercostals, and that the internal intercostals extend all the way from the dorsal end of the costal interspace to the sternum, whereas the external intercostals cover the dorsal, but not the ventral part of the interspace.

SACROSPINALIS The sacrospinalis (not illustrated) is the muscle which lies dorsal to the spinal column, extending from the pelvis to the head. It is invested by the lumbodorsal fascia, and originates from the dorsal aspect of the pelvis and spinal column. In the thoracic region it is divisible into several elements which insert on the ribs, the transverse process of the vertebrae, and the head. The actions of its subdivisions are complex and will not be described in detail, but in general it may be said that the sacrospinalis acts to extend the spinal column, draw the ribs posteriorly, and bend the neck and spinal column to one side.

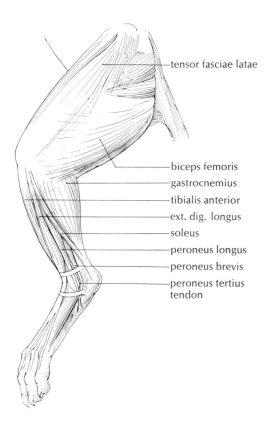

tensor fasciae latae

biceps femoris
gastrocnemius
tibialis anterior
ext. dig. longus
soleus
peroneus longus
peroneus brevis
peroneus tertius
tendon

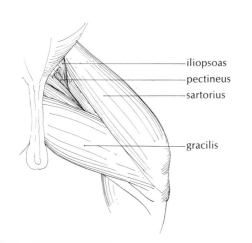

iliopsoas
pectineus
sartorius

gracilis

Clear the superficial fat and fascia away from the hip and thigh, being careful not to damage the spermatic cord if your specimen is a male. Examine the lateral aspect of the thigh and identify the muscles illustrated in Figure 24. Do not cut the fascia lata, which is the tough white aponeurosis on the anterior and lateral aspect of the thigh. The fascia lata is continuous proximally with the fascia of the gluteal muscles; distally it is attached to the ligaments of the patella and is continuous with the fascia of the lower limb.

The tensor fasciae latae is a thick triangular muscle which is continuous with the proximal portion of the fascia lata. It originates from the crest of the ilium and neighboring fascia and inserts on the fascia lata. It acts to tighten the fascia lata and to draw the thigh anteriorly.

The biceps femoris is the large muscle covering the lateral portion of the thigh. It originates from the tuberosity of the ischium and inserts on the patella, tibia, and fascia of the lower limb. It abducts the thigh and assists in flexing the knee.

The gastrocnemius is the large muscle on the posterior aspect of the lower hindlimb. It originates as lateral and medial heads from the distal end of the femur and from the fascia of the knee. It inserts on the calcaneus, and acts as an extensor of the foot.

The soleus lies deep to the gastrocnemius on the lateral side. It originates from the fibula and inserts on the calcaneus. Both the soleus and the gastrocnemius extend the foot.

The peroneus longus originates from the proximal portion of the fibula and inserts by a tendon which passes through a groove on the lateral malleolus and then turns medially to attach to the bases of the metacarpals. It extends the foot.

The peroneus brevis originates from the distal portion of the fibula and inserts on the base of the fifth metatarsal. It is an extensor of the foot.

The peroneus tertius is a slender muscle which lies between the peroneus longus and brevis. It originates from the lateral side of the fibula and inserts on the extensor tendon of the fifth digit. It assists in extending the fifth digit and in flexing the foot.

The extensor digitorum longus originates from the lateral epicondyle of the femur. It inserts on the bases of the middle and distal phalanges by means of four tendons which spread over the dorsum of the foot. It extends the digits.

The tibialis anterior originates from the proximal ends of the tibia and fibula and inserts on the first metatarsal. It flexes the foot.

Examine the medial aspect of the thigh and identify the muscles illustrated in Figure 25.

The sartorius covers the anterior half of the medial aspect of the thigh. It originates from the crest and ventral border of the ilium and inserts on the tibia, patella, and fascia of the knee. It adducts and rotates the thigh and extends the knee.

The gracilis covers the posterior portion of the medial aspect of the thigh. It originates from the ischium and the pubic symphysis and inserts by a broad aponeurosis on the medial surface of the tibia. It adducts the leg and draws it posteriorly.

The iliopsoas (homolog of the psoas major and iliacus of man) arises from the lumbar vertebrae and from the ilium. It inserts

FIG. 24.

SUPERFICIAL MUSCLES OF THE HINDLIMB, LATERAL VIEW

1	aponeurosis of tensor fasciae latae	10	peroneus brevis
2	biceps femoris	11	peroneus brevis tendon
3	caudofemoralis	12	peroneus longus
4	extensor digitorum longus	13	peroneus tertius
5	extensor digitorum longus tendons	14	peroneus tertius tendon
		15	sartorius
6	fascia lata	16	soleus
7	gastrocnemius	17	tensor fasciae latae
8	gluteus maximus	18	tibialis anterior
9	lateral malleolus	19	transverse ligaments
		20	semitendinosus

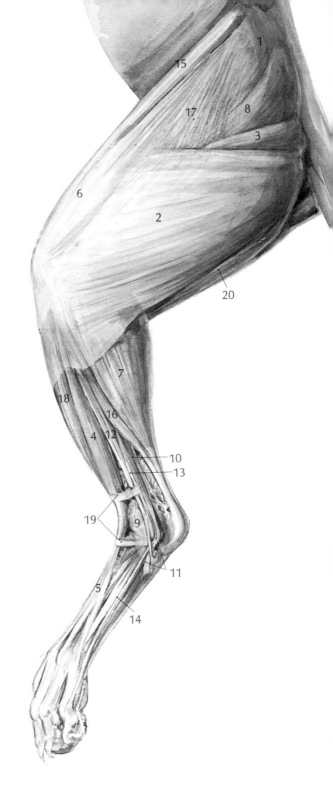

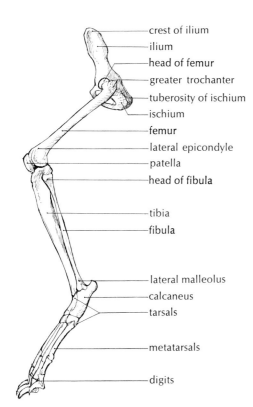

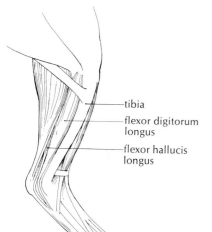

on the lesser trochanter of the femur, acting to draw the thigh forward and rotate it outward.

The pectineus originates from the anterior border of the pubis and inserts on the proximal end of the femur. It adducts the thigh.

The flexor digitorum longus lies on the medial aspect of the lower limb, next to the tibia. It arises from the proximal portion of the tibia, fibula, and adjacent fascia. It inserts by four tendons on the bases of the terminal phalanges and acts as a flexor of the digits.

The flexor hallucis longus originates lateral to the flexor digitorum longus. It inserts, in common with the flexor digitorum longus tendons, on the digits, and assists in flexing them.

31

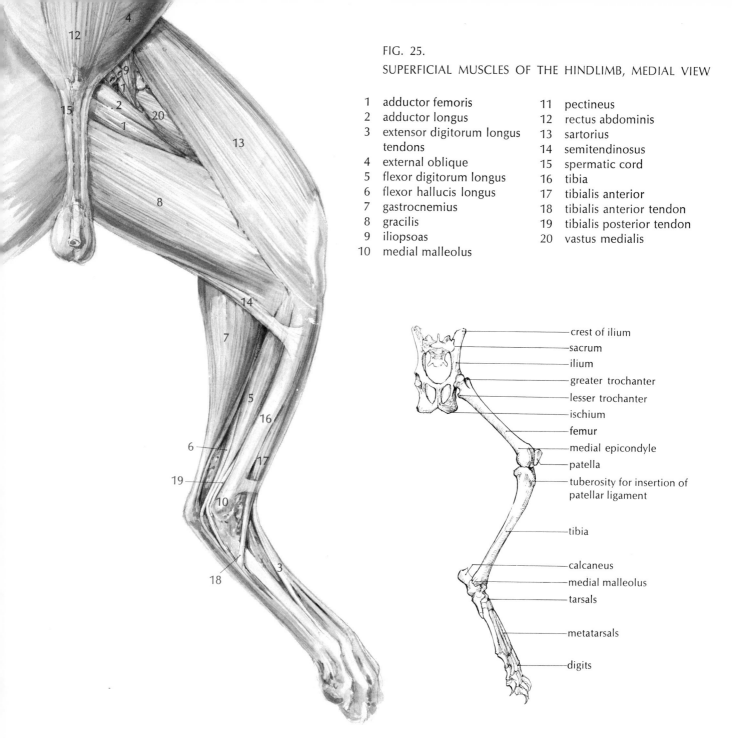

FIG. 25.

SUPERFICIAL MUSCLES OF THE HINDLIMB, MEDIAL VIEW

1	adductor femoris	11	pectineus
2	adductor longus	12	rectus abdominis
3	extensor digitorum longus tendons	13	sartorius
4	external oblique	14	semitendinosus
5	flexor digitorum longus	15	spermatic cord
6	flexor hallucis longus	16	tibia
7	gastrocnemius	17	tibialis anterior
8	gracilis	18	tibialis anterior tendon
9	iliopsoas	19	tibialis posterior tendon
10	medial malleolus	20	vastus medialis

crest of ilium
sacrum
ilium
greater trochanter
lesser trochanter
ischium
femur
medial epicondyle
patella
tuberosity for insertion of patellar ligament

tibia

calcaneus
medial malleolus
tarsals

metatarsals

digits

DISSECTION OF LATERAL THIGH MUSCLES

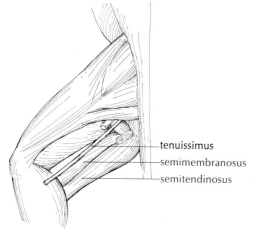

tenuissimus
semimembranosus
semitendinosus

Referring to Figures 24 and 26, remove the biceps femoris and expose the deep muscles on the lateral aspect of the thigh. Your dissection should resemble Figure 26. Deep to the biceps femoris you will find a large nerve, the sciatic (Fig. 71, p. 98), and near it a slender muscle, the tenuissimus. The tenuissimus originates from the second caudal vertebra and inserts in common with the biceps femoris. It has no homolog in man.

The semimembranosus originates from the ischium and inserts on the medial epicondyle of the femur and on the proximal end of the tibia. It draws the thigh posteriorly.

The semitendinosus lies on the posterior aspect of the thigh. It originates from the tuberosity of the ischium and inserts by a

FIG. 26.
DEEP MUSCLES OF THE HINDLIMB, DORSAL VIEW

1 adductor femoris
2 caudofemoralis
3 cut origin of biceps
4 extensor digitorum longus
5 fascia lata over vastus
   lateralis
6 gastrocnemius
7 gluteus maximus
8 gluteus medius
9 lumbodorsal fascia

10 peroneus brevis
11 peroneus brevis tendon
12 peroneus longus
13 peroneus tertius tendon
14 sciatic nerve
15 semimembranosus
16 semitendinosus
17 soleus
18 tensor fasciae latae
19 tenuissimus

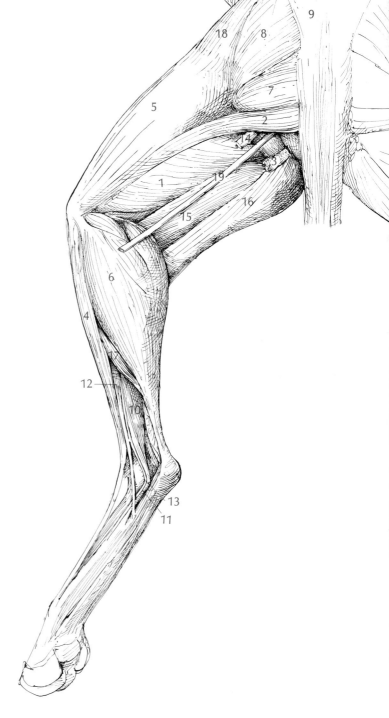

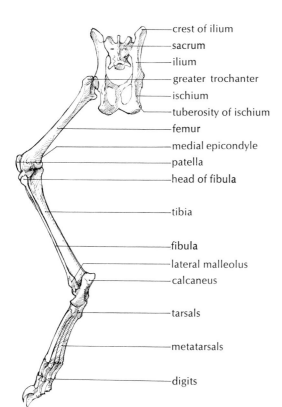

crest of ilium
sacrum
ilium
greater trochanter
ischium
tuberosity of ischium
femur
medial epicondyle
patella
head of fibula
tibia
fibula
lateral malleolus
calcaneus
tarsals
metatarsals
digits

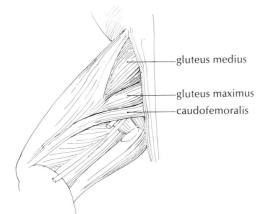

gluteus medius
gluteus maximus
caudofemoralis

thin tendon on the medial side of the tibia. It flexes the knee.
   The gluteus medius originates from the ilium and the transverse
processes of the last sacral and first caudal vertebrae. It inserts on the
greater trochanter of the femur and acts as an abductor of the thigh.
   The caudofemoralis originates from the transverse processes
of the second and third caudal vertebrae and inserts on the
patella and the surrounding fascia by a thin tendinous band.
It abducts the thigh and helps to extend the knee. It has no homolog
in man.
   The gluteus maximus lies between the gluteus medius and
the caudofemoralis. It originates from the transverse processes of the
last sacral and first caudal vertebrae and inserts on the greater
trochanter. It abducts the thigh.

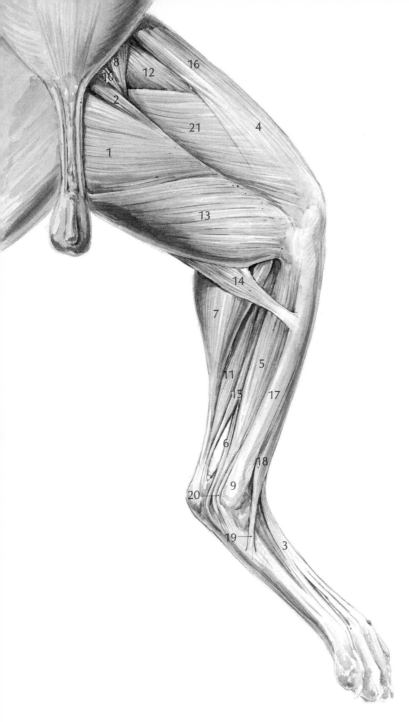

FIG. 27.
DEEP MUSCLES OF THE HINDLIMB, MEDIAL VIEW

1   adductor femoris
2   adductor longus
3   extensor digitorum longus
    tendons
4   fascia lata
5   flexor digitorum longus
6   flexor hallucis longus
7   gastrocnemius
8   iliopsoas
9   medial malleolus
10 pectineus
11 plantaris
12 rectus femoris
13 semimembranosus
14 semitendinosus
15 soleus
16 tensor fasciae latae
17 tibia
18 tibialis anterior
19 tibialis anterior tendon
20 tibialis posterior tendon
21 vastus medialis

DISSECTION OF MEDIAL THIGH MUSCLES

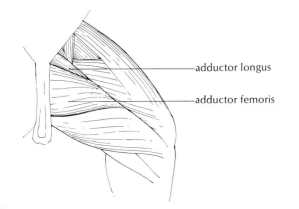

adductor longus

adductor femoris

Referring to Figures 25 and 27, remove the sartorius and the gracilis. Then, referring to Figures 27 and 28, remove the tensor fasciae latae. Identify the muscles illustrated in Figures 27 and 28.

The adductor longus and adductor femoris originate from the pubis and insert on the femur. They adduct the thigh.

The quadriceps femoris is the large muscle which covers the anterior surface of the thigh. It consists of four separate heads which are united distally by the patellar ligament. This ligament contains the patella and inserts on the proximal end of the tibia. The four heads of the quadriceps femoris are: the vastus lateralis, which originates from the lateral surface of the femur; the

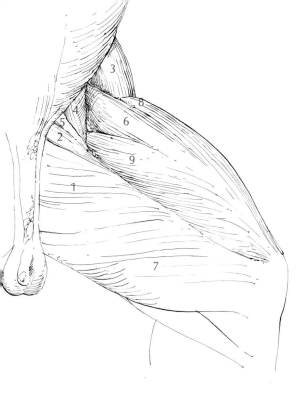

FIG. 28.

DEEP MUSCLES OF THE THIGH, MEDIAL VIEW

1   adductor femoris
2   adductor longus
3   gluteus medius
4   iliopsoas
5   pectineus
6   rectus femoris
7   semimembranosus
8   vastus lateralis
9   vastus medialis

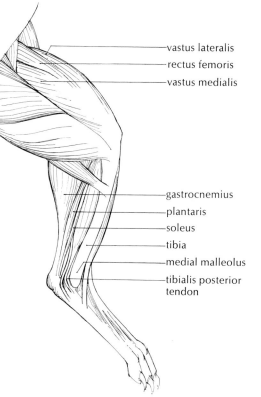

rectus femoris, which originates from the ilium anterior to the acetabulum; the vastus medialis, which originates from the medial side of the femur; and the vastus intermedius (not illustrated), which originates from the anterior surface of the femur. The vastus intermedius is the deep muscle next to the femur. It can be seen by separating the rectus femoris from the vastus lateralis. The four heads of the quadriceps femoris act together to extend the knee.

Separate the gastrocnemius from the soleus and identify the plantaris, which lies between the gastrocnemius and the soleus. The plantaris originates from the lateral side of the femur and the patella. It is fused with the lateral head of the gastrocnemius but is easily separated from the medial head. Its tendon passes around the calcaneus, lying within a sheath formed by the tendons of the soleus and gastrocnemius, and divides into four slips which insert on the bases of the second phalanges. The plantaris flexes the digits and acts together with the soleus and the gastrocnemius to extend the foot. In man the plantaris is a very slender muscle which inserts on the calcaneus.

The tibialis posterior originates from the proximal end of the tibia and fibula, and inserts on the medial side of the tarsals. It extends the foot.

Examine the abdominal muscles. The external oblique has already been observed in connection with the dissection of the chest and shoulder muscles. The most superficial of the muscles of the lateral abdominal wall, it originates from the lumbodorsal fascia and from the last nine or ten ribs by slips which interdigitate with the serratus ventralis and the serratus dorsalis. It inserts by a broad aponeurosis on the linea alba, or tendinous ventral midline of the abdominal wall.

Cut through the middle of the external oblique, making your cut at right angles to the direction of the fibers. Separate the external oblique from the internal oblique, which lies deep to it.

INTERNAL OBLIQUE   The internal oblique originates from the pelvis and the lumbodorsal fascia and inserts by a broad aponeurosis on the linea alba. Observe that the fibers of the internal oblique pass ventrally and anteriorly, whereas the fibers of the external oblique pass ventrally and posteriorly.

TRANSVERSUS   The transversus constitutes the third and deepest layer of the abdominal wall. It originates from the posterior ribs, lumbar vertebrae, and ilium, and inserts on the linea alba. The transversus is thin and may be difficult to distinguish from the internal oblique. The external oblique, internal oblique, and transversus act together to compress the abdominal viscera, as in defecation and in forced expiration, and to arch the back.

RECTUS ABDOMINIS   The rectus abdominis lies on the ventral aspect of the abdomen. It is enclosed in a tough sheath formed by the aponeuroses of the external oblique, internal oblique, and transversus. The rectus abdominis originates from the pubic symphysis and inserts on the sternum and costal cartilages. It acts with the other muscles of the abdominal wall to flex the trunk and to compress the abdominal viscera.

THE DIGESTIVE AND RESPIRATORY SYSTEMS

EXPOSURE OF SALIVARY GLANDS

The left parotid and submaxillary glands were identified and removed during the dissection of the muscles of the head and neck. A more detailed study of the salivary glands will now be made on the right side.

Remove the skin and platysma from the right side of the head and neck and identify the structures illustrated in Figure 29. The lymph nodes which lie on either side of the anterior facial vein should be removed; identify them by referring to Figures 15 and 16 (pp. 20 and 21). Distinguish them from the salivary glands by observing that the surfaces of the lymph nodes are smooth, whereas the surfaces of the salivary glands are lobulated.

The most prominent of the salivary glands is the parotid. Its duct crosses the masseter muscle and penetrates the cheek, opening into the mouth opposite the last cusp of the third upper premolar. To locate the opening of the parotid duct look inside the cheek and tug lightly on the duct with forceps.

Lift the anterior margin of the submaxillary gland and clear away the connective tissue to find the duct of the submaxillary gland. Near this duct is the small, oblong sublingual gland.

Cut the digastric and mylohyoid muscles and trace the duct of the submaxillary gland forward as far as possible in the floor of the mouth. It is accompanied by the duct of the sublingual gland, which is too small to be easily identified. The ducts of the sublingual and submaxillary glands open on a pair of small papillae at the base of the tongue.

The molar gland is a small, rather diffuse mass of glandular tissue located near the corner of the mouth between the masseter and the mandible. It opens into the mouth by several small ducts, too small to identify grossly.

Another salivary gland, the infraorbital gland, lies within the floor of the orbit and will be seen when the eye is dissected.

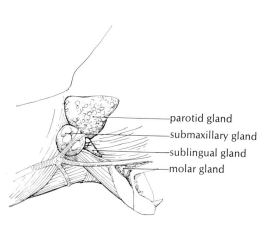

parotid gland
submaxillary gland
sublingual gland
molar gland

SAGITTAL SECTION OF HEAD AND NECK

The mouth, pharynx, and larynx can best be studied as seen in the sagittal section of the head and neck. The instructor should select certain students to make sagittal sections which may be studied by the entire class. Most specimens should be left intact for dissection during the study of the circulatory, nervous, and respiratory systems.

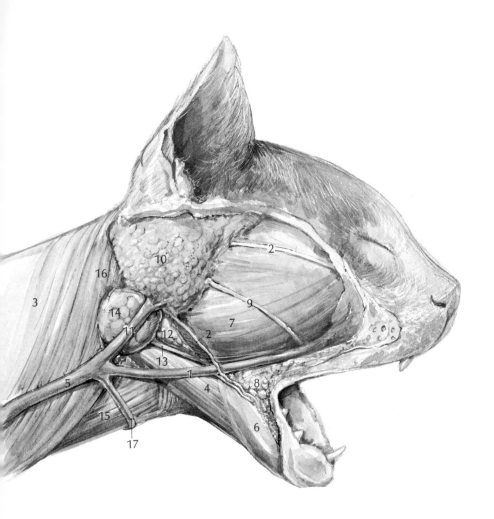

FIG. 29.
THE SALIVARY GLANDS

1 anterior facial vein
2 branch of facial nerve
3 clavotrapezius
4 digastric
5 external jugular vein
6 mandible
7 masseter
8 molar gland
9 parotid duct
10 parotid gland
11 posterior facial vein
12 sublingual gland
13 submaxillary duct
14 submaxillary gland
15 sternohyoid
16 sternomastoid
17 transverse jugular vein

SAGITTAL SECTION OF HEAD AND NECK

Use a bone saw to cut the head and neck in the sagittal plane. Wash the sections and identify the structures illustrated in Figure 30. Also refer to a skull cut in the sagittal plane and to Figure 5 on page 7. Review the names of the bones seen in this section and observe the relations of the bones to the other structures of the head and neck.

TEETH

For convenience the teeth may be described by the dental formula:

$$3 - 1 - 3 - 1$$
$$3 - 1 - 2 - 1$$

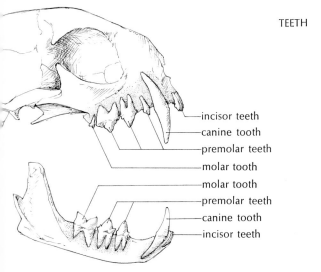

incisor teeth
canine tooth
premolar teeth
molar tooth
molar tooth
premolar teeth
canine tooth
incisor teeth

The numbers represent, from left to right, the number of incisors, canines, premolars and molars on either side. Figures in the upper row represent teeth in the upper jaw; figures in the lower row represent teeth in the lower jaw. In the cat the deciduous teeth appear two or three weeks after birth, and are replaced by permanent teeth at about seven months. Observe that the incisors of both jaws and the first premolar and molar of the upper jaw are quite small compared with the other teeth. Also observe that the last upper molar and the lower molar form a shearing mechanism, typical of carnivores, as opposed to the grinding mechanism found in herbivores.

TONGUE

Observe the papillae of the tongue. Those in the central portion carry small spines which serve as scrapers. Similar spines may be seen on the prominent transverse ridges of the hard palate. At the sides and back of the tongue are softer and larger papillae.

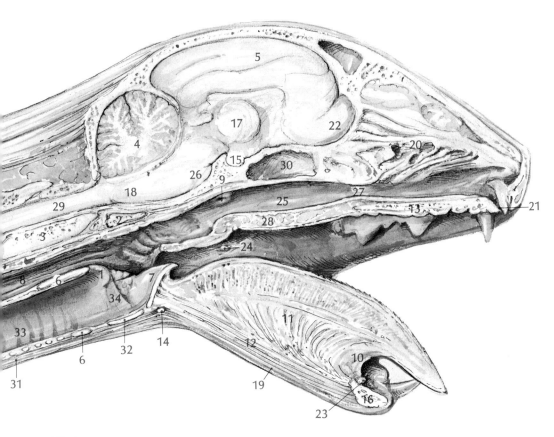

FIG. 30.

SAGITTAL SECTION OF THE HEAD AND NECK

1	arytenoid cartilage	14	hyoid bone	25	pharynx
2	atlas	15	hypophysis	26	pons
3	axis	16	mandible	27	posterior naris (choana)
4	cerebellum	17	massa intermedia	28	soft palate
5	cerebrum	18	medulla	29	spinal cord
6	cricoid cartilage	19	mylohyoid	30	sphenoid sinus
7	epiglottis	20	nasal conchae	31	sternohyoid
8	esophagus	21	nasopalatine duct	32	thyroid cartilage
9	Eustachian tube	22	olfactory bulb	33	trachea
10	frenulum	23	openings of submaxillary	34	vocal cord
11	genioglossus		and sublingual gland		
12	geniohyoid		ducts		
13	hard palate	24	palatine tonsil		

GENIOGLOSSUS   The genioglossus is the largest muscle of the tongue, composing most of its bulk as seen in sagittal section. This muscle originates from the anterior end of the mandible. Its fibers pass back in a fan-shaped pattern to insert along the tip, dorsum, and root of the tongue and on the hyoid bone. The various parts of the genioglossus have different actions. The posterior part protrudes the tongue, the middle part depresses the tongue, and the anterior part retracts the tip of the tongue. In addition to the genioglossus there are two other tongue muscles, the styloglossus and the hyoglossus (not seen in sagittal section), which originate on the hyoid bone and insert on the tongue. There are also several systems of muscle fibers which have both their origins and insertions within the tongue.

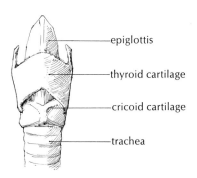

epiglottis
thyroid cartilage
cricoid cartilage
trachea

larynx, ventral view

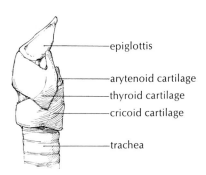

epiglottis
arytenoid cartilage
thyroid cartilage
cricoid cartilage
trachea

larynx, lateral view

TONSILS The palatine tonsils are small rounded masses of lymphatic tissue located in the lateral wall of the soft palate near the base of the tongue. The visible portion, which projects into the mouth, is a relatively small part of the tonsil, most of which is embedded in the mucous membrane.

HARD PALATE The anterior portion of the mouth is separated from the nasal conchae by the hard palate. Refer to Figures 5 and 6 on pages 7 and 8 and review the bones of the hard palate.

PHARYNX The pharynx is the cavity dorsal to the soft palate and the larynx. Anteriorly it communicates with the nasal cavities via the internal nares. Laterally the Eustachian tubes open into it, and posteriorly it communicates with the esophagus, larynx, and mouth. The free posterior border of the soft palate forms an opening termed the isthmus of the fauces, which leads from the mouth to the pharynx. Normally this opening is closed by the base of the tongue. It opens only during swallowing or in breathing through the mouth. Within the walls of the pharynx are a number of muscles which act to increase the transverse diameter of the pharnyx as food enters it and then contract upon the food, sending it down into the esophagus.

LARYNX The larynx is situated at the anterior end of the trachea, forming a passageway between the trachea and the pharnyx. It consists of a framework of cartilages connected by ligaments, muscles, and membranes. The cartilages of the larynx and their relation to the hyoid bone should be seen in a demonstration dissection made by the instructor. Leave them intact in your specimen in order to preserve the vessels and nerves lateral to the larynx.

Most of the lateral and ventral walls of the larynx are supported by the large, shield-shaped thyroid cartilage. Posterior to the thyroid cartilage is the ring-shaped cricoid cartilage, which is much wider dorsally than it is ventrally. Anterior to the dorsal part of the cricoid cartilage, the dorsolateral rim of the glottis is supported by the small paired arytenoid cartilages. The vocal cords are paired folds of the mucosa between the arytenoid cartilages and the thyroid cartilage. The epiglottis is supported by the epiglottic cartilage, which is attached to the midventral part of the anterior border of the thyroid cartilage. When food is swallowed, the larynx is raised and the epiglottis folds over the glottis, or entrance to the trachea. The epiglottis guides the food into the esophagus and prevents it from entering the trachea.

THYROID GLAND The thyroid gland consists of paired lobes which lie on either side of the trachea just posterior to the cricoid cartilage. They are connected by a thin strip of thyroid tissue, termed the isthmus, which extends across the ventral side of the trachea. See the thyroid gland as illustrated in Figure 47, page 67.

TRACHEA The trachea is composed of a fibrous membrane stiffened by cartilaginous rings, which serve to keep the air passage open. The dorsal side of the trachea is in contact with the esophagus; in this area the cartilaginous rings are deficient, and the tracheal tube is completed by fibrous tissue and smooth muscle fibers.

EXPOSURE OF LUNG Use bone scissors to make a cut parallel to, and about an inch lateral to, the left side of the sternum. Cut through the muscles and costal cartilages, extending the cut from the first rib to the posterior end of the sternum. Then extend the anterior and

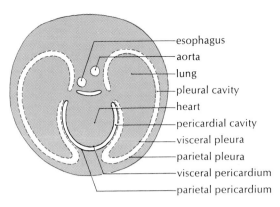

- esophagus
- aorta
- lung
- pleural cavity
- heart
- pericardial cavity
- visceral pleura
- parietal pleura
- visceral pericardium
- parietal pericardium

cross section of thorax

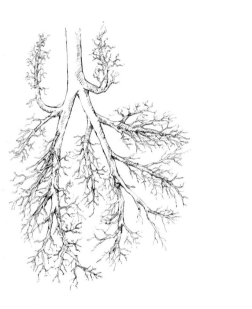

cast of right bronchus, lateral view

posterior ends of the cut dorsally between adjacent ribs and pull back the thoracic wall to expose the left lung.

The potential space between the lung and the thoracic wall is the pleural cavity. There are two pleural cavities, right and left, which are completely separated from each other. Observe the delicate pleural fold which passes from the heart to the ventral median line. The space between the pleural cavities is termed the mediastinum. It extends from the sternum to the vertebral column and contains all the thoracic viscera except the lungs and pleurae.

The pleura is the thin, serous membrane which encloses each pleural cavity. That portion of the pleura which covers the lungs is termed the visceral pleura. The portion that covers the inner surface of the chest wall, the diaphragm, and the structures of the mediastinum is termed the parietal pleura. The visceral and parietal pleurae are continuous with each other at the root of the lung and normally lie in direct contact with each other. In life they are moistened by a serous secretion which allows the lungs to move freely against the chest wall.

Cut the abdominal and thoracic walls to make a dissection similar to Figure 31. Be careful not to damage the brachial nerves and vessels; cut only through the body wall, leaving all other structures intact.

LUNGS

Examine the lungs, and observe that each lung is divided into three lobes: anterior, middle, and posterior. The posterior lobe of the right lung is subdivided into medial and lateral portions. About the level of the sixth rib the trachea divides into two branches termed right and left main bronchi. Each main bronchus divides within the substance of the lung, sending a large branch to each lobe. These branches subdivide into successively smaller branches which finally terminate in alveoli, minute air sacs which are richly supplied with capillaries.

PERICARDIUM

The heart is enclosed in a membranous sac termed the pericardium. The pericardium consists of two parts: the visceral pericardium is closely adherent to the heart wall; the parietal pericardium is continuous with the visceral pericardium around the roots of the great vessels and forms a sac which encloses the heart. The potential space between the parietal pericardium and the visceral pericardium is termed the pericardial cavity. In life it contains a serous fluid which allows the heart to move freely against the parietal pericardium.

THYMUS GLAND

Identify the thymus gland, a mass of glandular tissue lying in the median ventral line near the anterior end of the heart. It is prominent in young animals, but may be difficult to identify in older specimens.

ROOT OF LUNG

Cut open the pericardium and examine its relation to the heart. Press the heart away from the lung on one side and examine the place where the blood vessels and main bronchus enter the lung. This is termed the root of the lung. Pull the lung forward and examine the pulmonary ligament, a double fold of pleura which attaches each lung to the aorta, vertebral column, and diaphragm.

EXAMINATION OF ABDOMINAL VISCERA

Lift the stomach and liver and explore the contents of the abdominal cavity. Identify the structures you encounter by referring to the illustrations in this section, but do not cut any structures until instructed to do so and do not attempt to trace the vessels and

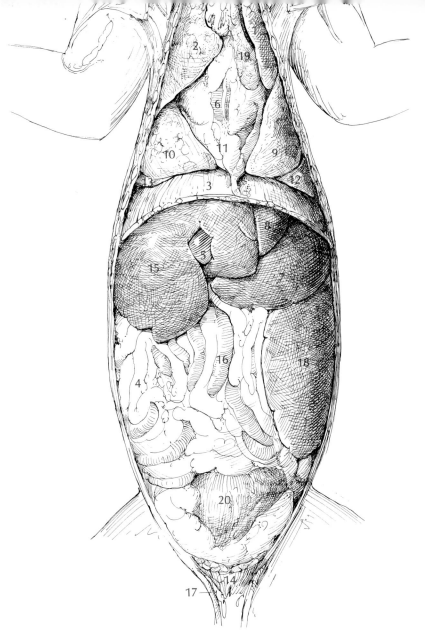

FIG. 31.

THE ABDOMINAL AND THORACIC
VISCERA, VENTRAL VIEW

1 anterior lobe of left lung
2 anterior lobe of right lung
3 diaphragm
4 fat in omentum
5 gallbladder
6 heart within pericardium
7 left lateral lobe of liver
8 left medial lobe of liver
9 middle lobe of left lung
10 middle lobe of right lung
11 pericardial fat
12 posterior lobe of left lung
13 posterior lobe of right lung
14 rectus abdominis
15 right medial lobe of liver
16 small intestine covered by
     omentum
17 spermatic cord
18 spleen
19 thymus gland
20 urinary bladder

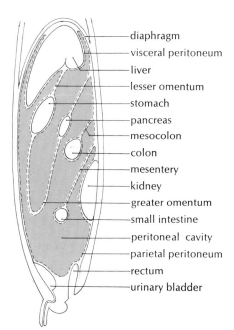

diaphragm
visceral peritoneum
liver
lesser omentum
stomach
pancreas
mesocolon
colon
mesentery
kidney
greater omentum
small intestine
peritoneal cavity
parietal peritoneum
rectum
urinary bladder

sagittal section of abdomen

ducts at this time.

Both the inside of the body wall and the viscera are lined by a serous membrane, the peritoneum. That portion of the peritoneum which lines the viscera is the visceral peritoneum, and that portion of the peritoneum which lines the abdominal wall is the parietal peritoneum. The peritoneal cavity is the potential space between the parietal and visceral layers. In life it contains a serous fluid which allows the viscera to glide upon each other and upon the body wall with minimal friction. The peritoneum may be likened to a closed sac into which the abdominal organs have been pushed from the outside. It covers the organs, but none of them actually lie within the peritoneal cavity.

Structures such as the kidneys, which are attached to the body wall and covered by peritoneum on only one side, are said to be retroperitoneal, or behind the peritoneum. The fact that the ventral surface of the kidney is covered by peritoneum can easily be demonstrated by separating the peritoneum from the kidney with forceps and small scissors. In other areas the peritoneum adheres so closely to the organs which it covers that it cannot be dissected away from them. The peritoneal coverings of the liver and spleen, for instance, cannot be demonstrated by gross dissection.

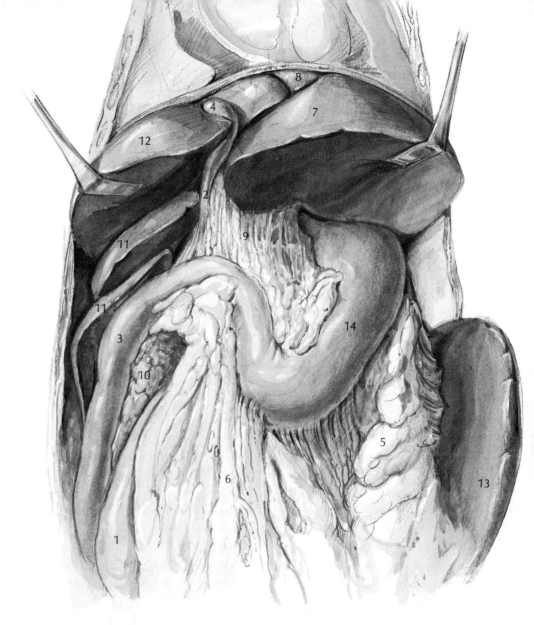

MESENTERIES, LIGAMENTS, AND OMENTA

The peritoneal sheets which extend between the body wall and the viscera are termed mesenteries, ligaments, and omenta. Within these sheets are the vessels, nerves, and lymphatics which supply the viscera. Examine the greater omentum. It is a double sheet of peritoneum which is attached to the greater curvature of the stomach and to the dorsal body wall. Manipulate the greater omentum and determine that it is composed of dorsal and ventral layers, with a potential space, the omental bursa, between them. Lift the greater omentum and observe its relations to the pancreas and the spleen. The dorsal layer of the greater omentum encloses the gastrosplenic part of the pancreas. That portion of the greater omentum between the stomach and the spleen is termed the gastrosplenic ligament.

The stomach and duodenum are attached to the liver by the lesser omentum, which consists of two portions: the gastrohepatic ligament, between the stomach and the liver, and the hepatoduodenal ligament, between the liver and the duodenum. Within the right lateral border of the lesser omentum are the common bile duct, the hepatic artery, and the portal vein.

Make a slit in the greater omentum and confirm the fact that

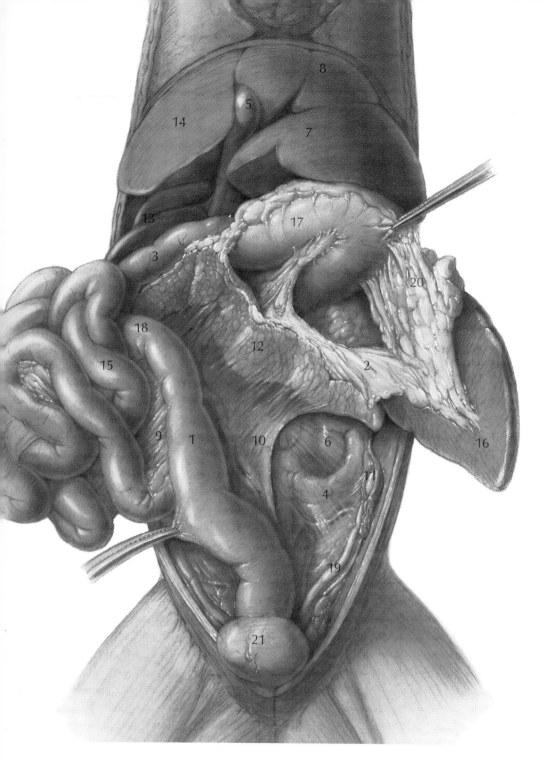

FIG. 33.
THE ABDOMINAL VISCERA,
VENTRAL VIEW

The omentum is removed and
the large and small intestines
are retracted to the right.

1 descending colon
2 dorsal layer of gastrosplenic ligament
3 duodenum
4 fat (retroperitoneal)
5 gallbladder
6 kidney
7 left lateral lobe of liver
8 left medial lobe of liver
9 mesentery
10 mesocolon
11 ovary
12 pancreas
13 right lateral lobe of liver
14 right medial lobe of liver
15 small intestine
16 spleen
17 stomach
18 transverse colon
19 uterine horn
20 ventral layer of gastro-
    splenic ligament
21 urinary bladder

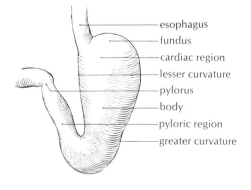

esophagus
fundus
cardiac region
lesser curvature
pylorus
body
pyloric region
greater curvature

it consists of dorsal and ventral layers which may be separated.

Trim away the greater omentum, leaving the gastrosplenic
ligament, to make a dissection similar to Figure 33.
Pass a probe dorsal to the right lateral border of the lesser omentum
and observe that there is a passage between the peritoneal cavity
and the omental bursa. This passage is termed the epiploic foramen.

The opening by which the stomach communicates with the
esophagus is termed the cardiac orifice, and the opening by which the
stomach communicates with the duodenum is termed the pylorus.
The pylorus is surrounded by a muscular ring, the pyloric valve,
which regulates the passage of food from the stomach to the
duodenum.

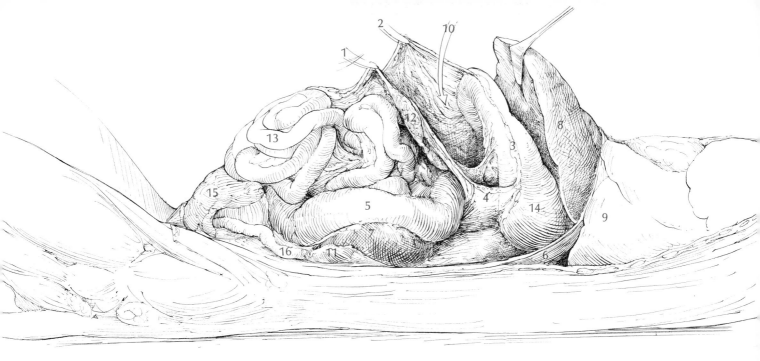

FIG. 34.
THE ABDOMINAL VISCERA, LATERAL VIEW

The spleen and gastrosplenic ligament are removed.

1	clamp on dorsal layer of greater omentum	5	descending colon	12	pancreas within dorsal layer of greater omentum
2	clamp on ventral layer of greater omentum	6	diaphragm	13	small intestine
3	cut attachment of greater omentum	7	kidney	14	stomach
		8	liver	15	urinary bladder
4	cut gastrosplenic ligament	9	lung	16	uterine horn
		10	omental bursa		
		11	ovary		

DUODENUM   Trace the small intestine distal to the pylorus. The first portion of the small intestine is the duodenum. It passes posteriorly for about three and a half inches and then doubles back on itself, forming a loop within which the duodenal portion of the pancreas lies. Examine the mesentery which supports the duodenum. It is

MESODUODENUM   termed the mesoduodenum, and encloses the duodenal portion of the pancreas. Referring to Figures 35 and 36, pick away portions of the lesser omentum as necessary to identify the common bile duct and trace it to the point where it enters the duodenum. The total length of the duodenum is about seven inches.

Spread out the mesentery of the small intestine, observing that it contains fat, lymph nodes, and vessels (the intestinal branches of the superior mesenteric artery and vein). The small

JEJUNUM AND ILEUM   intestine beyond the duodenum is divided into the jejunum (proximal half) and the ileum (distal half). This division is somewhat arbitrary, as there is no definite point of demarcation between jejunum and ileum.

LARGE INTESTINE   Trace the ileum to its junction with the ascending colon. The alimentary canal distal to this point is termed the large intestine. Identify the cecum and the ascending, transverse, and descending parts

FIG. 35.
VISCERAL SURFACE
OF THE LIVER

1 caudate lobe
2 coronary ligament
3 left medial lobe
4 nonperitoneal surface
5 postcava
6 right lateral lobe,
   anterior part
7 right lateral lobe,
   posterior part
8 right medial lobe

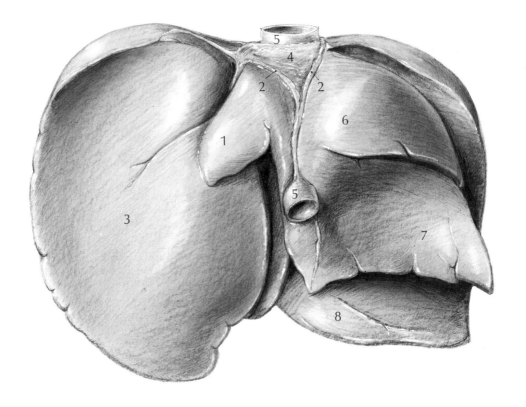

LIVER

of the colon as illustrated in Figure 36. The rectum is that portion of the large intestine which extends from the pelvic inlet to the anus. The mesentery of the colon is termed the mesocolon.

Refer to Figures 36 and 38 and identify the lobes of the liver. Identify the gallbladder and trace its connection with the hepatic, cystic, and common bile ducts. The form of the hepatic ducts, which convey bile from the lobes of the liver, varies. They may join the cystic duct by a common stem or by two or more separate stems.

The portal vein enters the liver and breaks up into a system of capillaries. Blood from the capillaries is returned to the vena cava via the hepatic veins. Note that the postcava passes through the liver, within which it receives the hepatic veins.

Refer to Figures 35, 36, and 37, and observe the relations of the common bile duct, the hepatic artery, and the portal vein. Dissect the lesser omentum away from these structures as necessary to expose them. Identify the falciform ligament, which lies in the cleft between the right and left medial lobes of the liver. In its free margin is the round ligament, which represents the obliterated umbilical vein. The falciform ligament is continuous with the coronary ligaments, which attach the liver to the central tendon of the diaphragm.

right medial lobe
round ligament
falciform ligament
hepatic veins opening into postcava
coronary ligament
right lateral lobe
left medial lobe
left lateral lobe

liver, anterior view

46

DIAPHRAGM   Examine the diaphragm in Figure 39 and observe that it consists of a central tendon surrounded by muscle fibers which originate from the costal cartilages, sternum, vertebrae, and fascia of the dorsal body wall. The diaphragm forms a partition between the thoracic and abdominal cavities. It contains openings for the passage of the esophagus, postcava, and aorta. The diaphragm is an important muscle of respiration. When it contracts, the central tendon is drawn posteriorly and the volume of the thoracic cavity is increased. The diaphragm also acts with the abdominal muscles to compress the contents of the abdominal cavity.

DISSECTION OF ILEUM AND JEJUNUM   Remove the ileum and jejunum, leaving a short piece of ileum attached to the colon as illustrated in Figure 36. Slit open several sections of small intestine and wash out the contents. Use a hand lens to observe the villi, minute tubular projections which serve to increase the mucosal surface.

Trim away the fat, lymph nodes, and connective tissue as necessary to make a dissection similar to Figure 36. Identify the structures illustrated.

The celiac artery, superior mesenteric artery, and portal vein will be studied at this time because of their close association with the alimentary canal. Read the following descriptions of these vessels, study Figures 36, 37, and 38, and trace the vessels as far as possible.

The celiac artery is the first vessel given off by the abdominal aorta. It divides into three branches: the hepatic, left gastric, and splenic arteries.

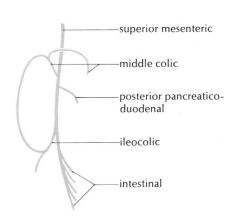
- cystic
- hepatic
- gastroduodenal
- hepatic
- celiac
- left gastric
- splenic
- pyloric
- right gastroepiploic
- ant. pancreaticodu-
  odenal

- superior mesenteric
- middle colic
- posterior pancreatico-
  duodenal
- ileocolic
- intestinal

The hepatic artery lies along the cranial border of the gastrosplenic part of the pancreas. Near the pylorus it turns cranially, lying in a fibrous sheath together with the portal vein and the common bile duct. It gives a cystic artery to the gallbladder and then branches to the lobes of the liver. The largest branch of the hepatic is the gastroduodenal artery, which it gives off near the pylorus. The gastroduodenal is a short vessel that gives rise to three branches: the pyloric artery (to the pylorus and lesser curvature of the stomach), the anterior pancreaticoduodenal artery (to the duodenum and pancreas), and the right gastroepiploic artery (to the greater omentum and the greater curvature of the stomach).

The left gastric artery lies along the lesser curvature of the stomach, supplying many branches to both dorsal and ventral stomach walls. It anastomoses with the pyloric artery.

The splenic artery is the largest branch of the celiac. It gives two or more branches to the dorsal surface of the stomach and then divides into anterior and posterior branches which supply the anterior and posterior ends of the spleen. The posterior branch also supplies a branch to the pancreas. Both anterior and posterior branches give several short gastric arteries to the greater curvature of the stomach.

Just below the celiac artery the abdominal aorta gives off the superior mesenteric artery. Its branches are: the posterior pancreaticoduodenal (to the caudal portions of the pancreas and duodenum), the middle colic (to the transverse and descending colon), the ileocolic (to the cecum and ileum). It may also give a separate right colic artery to the ascending colon. The superior mesenteric then divides into numerous intestinal branches

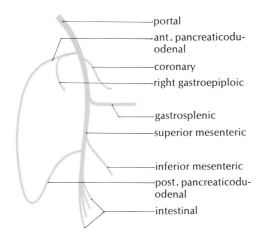

portal
ant. pancreaticodu-
odenal
coronary
right gastroepiploic

gastrosplenic
superior mesenteric

inferior mesenteric
post. pancreaticodu-
odenal
intestinal

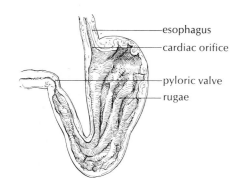

esophagus
cardiac orifice

pyloric valve
rugae

which supply the small intestine.

Blood from the celiac and superior mesenteric arteries passes through the capillaries of the alimentary canal, spleen, and pancreas, and then returns to the liver via the portal vein. Within the liver the portal vein ramifies, ending in a system of capillaries termed the sinusoids of the liver. From the sinusoids the blood passes to the postcava via the hepatic veins, which join the postcava within the substance of the liver.

The portal vein is formed by the union of the superior mesenteric and gastrosplenic veins. Near this union the portal vein is joined by the right gastroepiploic vein (from the greater curvature of the stomach), the anterior pancreaticoduodenal vein (from the duodenum and pancreas), and the coronary vein (from the lesser curvature of the stomach).

The gastrosplenic vein receives tributaries from the pancreas and stomach. It is formed by the union of two large branches termed anterior and posterior splenic veins, which accompany the two branches of the splenic artery to either end of the spleen.

The superior mesenteric vein is the largest tributary of the portal vein. It receives the posterior pancreaticoduodenal vein from the duodenum and pancreas, the inferior mesenteric vein from the descending colon and rectum, and numerous intestinal veins from the small intestine.

Remove the colon and mesocolon. Cut the stomach below the fundus, leaving the left gastric artery intact as shown in Figure 37. Cut the duodenum just proximal to the point where the common bile duct and the pancreatic duct join it, and remove the section of stomach and duodenum between the cuts. Open the section of stomach you have removed and observe the pyloric valve and the prominent folds of mucosa termed rugae. Cut open the colon and the ileum at the point where they join and observe the ileocecal valve, formed by a fold of ileum projecting into the cecum. Note that there are no villi in the colon.

Examine the pancreas. It is flattened, irregular, and quite variable in outline, and may be recognized by its lobulated texture. It consists of two parts, which lie at right angles to each other. The gastrosplenic part lies in the dorsal layer of the greater omentum, near the greater curvature of the stomach. The duodenal part lies in the mesoduodenum, near the medial border of the descending limb of the duodenum.

Starting at the point where the common bile duct joins the duodenum, pick away the pancreatic tissue and trace the pancreatic ducts as illustrated in Figure 37. The ampulla of Vater is the duct formed by the union of the common bile duct and the pancreatic duct at the point where they empty into the duodenum.

Identify the branches of the celiac artery, superior mesenteric artery, and portal vein as illustrated in Figure 37.

Remove the pancreas and duodenum, making a dissection similar to Figure 38. Identify the structures illustrated.

Remove the spleen and the liver to make a dissection similar to Figure 39. If you have a specimen in which the veins are injected with latex, try to dissect the liver away from the vena cava and preserve it as illustrated. If not, cut the vena cava and remove the liver intact.

FIG. 36.
THE STOMACH AND COLON

The jejunum and ileum are removed

1 anterior pancreatico-
duodenal artery
2 aorta
3 ascending colon
4 caudate lobe of liver
5 cecum
6 celiac artery
7 common bile duct
8 cystic duct
9 descending colon
10 descending limb of
duodenum
11 dorsal layer of gastrosplenic ligament
12 gastroduodenal artery
13 hepatic artery
14 hepatic duct
15 ileocolic artery
16 ileum
17 intestinal branches of
superior mesenteric a.
18 jejunum
19 left gastric artery
20 left lateral lobe of liver
21 lymph node
22 mesentery
23 mesocolon
24 middle colic artery
25 pancreas (duodenal part)
26 pancreas (gastrosplenic
part)
27 portal vein
28 pyloric artery
29 right gastroepiploic artery
30 right lateral lobe of liver
(anterior part)
31 right lateral lobe of liver
(posterior part)
32 right medial lobe of liver
33 splenic artery
34 spleen
35 stomach
36 superior mesenteric artery
37 superior mesenteric vein
38 transverse colon
39 ventral layer of gastrosplenic ligament
40 ventral layer of
greater omentum

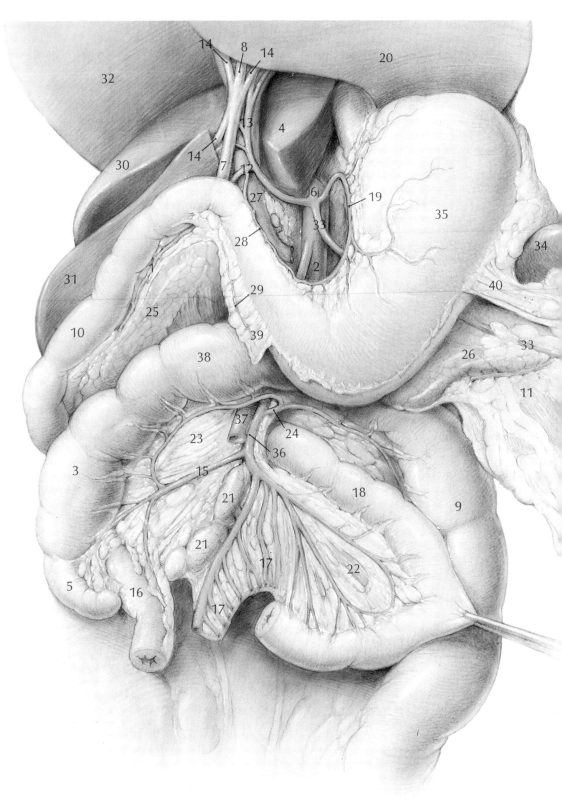

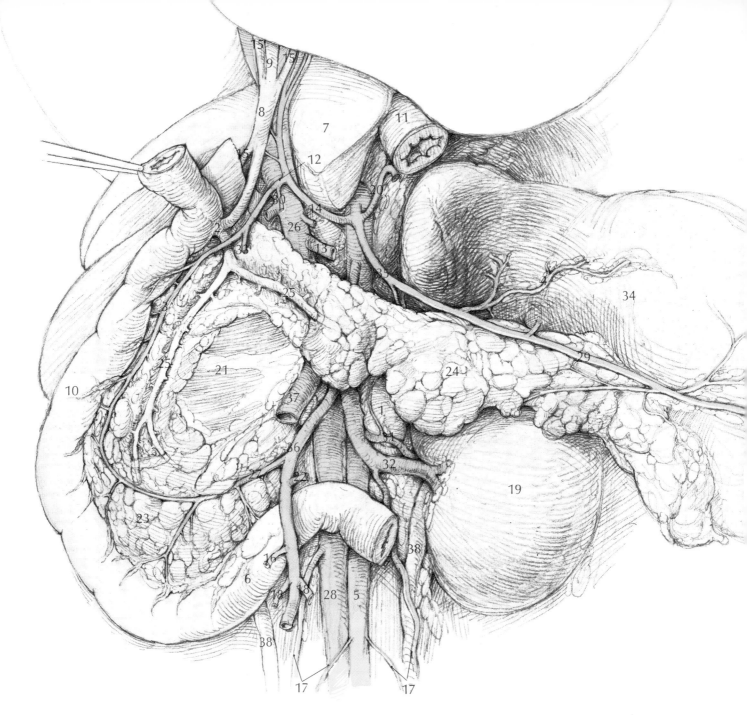

FIG. 37.
THE PANCREAS AND DUODENUM

The colon is removed and the pancreas is
dissected to expose the pancreatic ducts.

1	adrenal gland	11	esophagus	22	middle colic artery	33	right gastroepiploic
2	ampulla of Vater	12	gastroduodenal artery	23	pancreas (duodenal part)		artery
3	anterior pancreatico-duodenal artery	13	gastrosplenic vein	24	pancreas (gastrosplenic part)	34	spleen
4	anterior splenic artery	14	hepatic artery	25	pancreatic duct	35	splenic artery
5	aorta	15	hepatic duct	26	portal vein	36	superior mesenteric artery
6	ascending limb of duodenum	16	ileocolic artery	27	posterior pancreatico-duodenal artery	37	superior mesenteric vein
7	caudate lobe of liver	17	internal spermatic artery and vein	28	postcava	38	ureter
8	common bile duct	18	intestinal branch of superior mesenteric artery	29	posterior splenic artery		
9	cystic duct	19	kidney	30	pyloric artery		
10	descending limb of duodenum	20	left gastric artery	31	renal artery		
		21	mesoduodenum	32	renal vein		

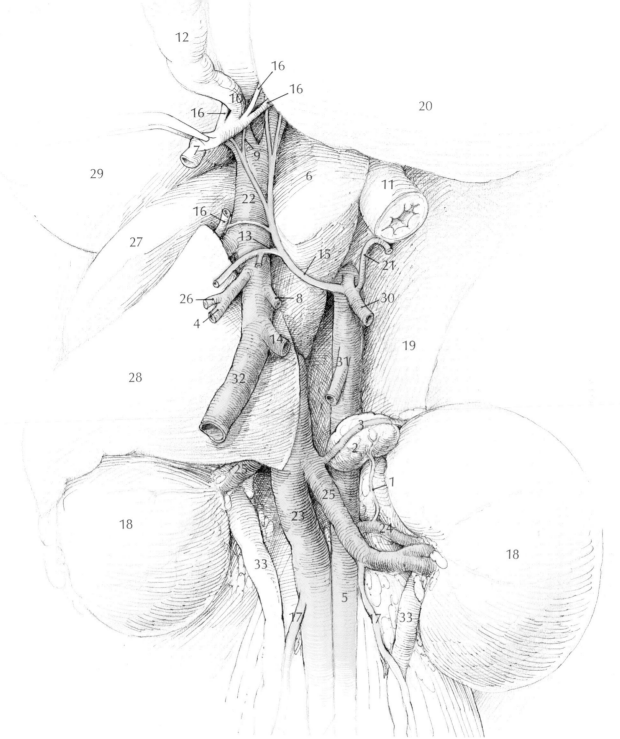

FIG. 38.
THE PORTAL VEIN AND CELIAC ARTERY

The pancreas and spleen are removed.

1	adrenal artery	12	gallbladder	24	renal artery
2	adrenal gland	13	gastroduodenal artery	25	renal vein
3	adrenolumbar vein	14	gastrosplenic vein	26	right gastroepiploic vein
4	anterior pancreatico-duodenal vein	15	hepatic artery	27	right lateral lobe of liver, anterior part
5	aorta	16	hepatic duct	28	right lateral lobe of liver, posterior part
6	caudate lobe of liver	17	internal spermatic v.	29	right medial lobe of liver
7	common bile duct	18	kidney	30	splenic artery
8	coronary vein	19	left crus of diaphragm	31	superior mesenteric a.
9	cystic artery	20	left lateral lobe of liver	32	superior mesenteric v.
10	cystic duct	21	left gastric artery	33	ureter
11	esophagus	22	portal vein		
		23	postcava		

THE UROGENITAL SYSTEM

EXPOSURE OF KIDNEYS AND URETERS

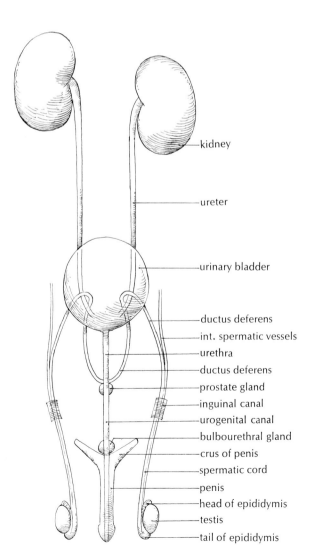

kidney

ureter

urinary bladder

ductus deferens
int. spermatic vessels
urethra
ductus deferens
prostate gland
inguinal canal
urogenital canal
bulbourethral gland
crus of penis
spermatic cord
penis
head of epididymis
testis
tail of epididymis

After removing the liver and spleen, clear away fat, peritoneum, and connective tissue as necessary to identify the structures illustrated in Figure 39 if your specimen is a male, or in Figure 42 if your specimen is a female. Special care should be taken to preserve the internal spermatic vessels and the ureter, which are surrounded by fat and may be accidentally cut. Review the branches of the abdominal aorta and postcava as described on page 68.

The kidneys are retroperitoneal; they are situated on either side of the vertebral column about the level of the third to fifth lumbar vertebrae, and are surrounded by fat. The right kidney normally lies somewhat anterior to the left kidney. The ductless adrenal gland lies near the anterior end of the kidney.

The kidney is enclosed in a fibrous renal capsule which may be separated easily from its surface. Strip off the capsule, observing its attachment to the ureter and renal vessels. The central depression in the medial surface of the kidney is termed the hilus; the renal vessels and the ureter join the kidney at this point. Trace the ureters to the bladder, observing their relations to the uterine horns (in the female) and to the ductus deferens (in the male).

Remove both kidneys. Use a razor blade to make a series of thin parallel longitudinal sections of the left kidney, continuing until you reach the renal papilla. Arrange the sections in serial order. The kidney should resemble Figure 40.

Similarly, make a series of cross sections of the right kidney, identifying the structures illustrated in Figure 41.

The substance of the kidney consists of a superficial cortex and a deeper, darker-colored medulla. The central cavity of the kidney is the renal sinus. It contains fat, branches of the renal vessels, and the renal pelvis, a cup-shaped extension of the anterior end of the ureter. The renal papilla is a cone-shaped projection enclosed by the pelvis.

Near the hilus, the renal artery and vein divide into dorsal and ventral branches. Within the kidney each branch divides into five or more interlobar arteries and veins. Near the dividing line between cortex and medulla the interlobar arteries and veins are connected with each other by the arcuate arteries and veins.

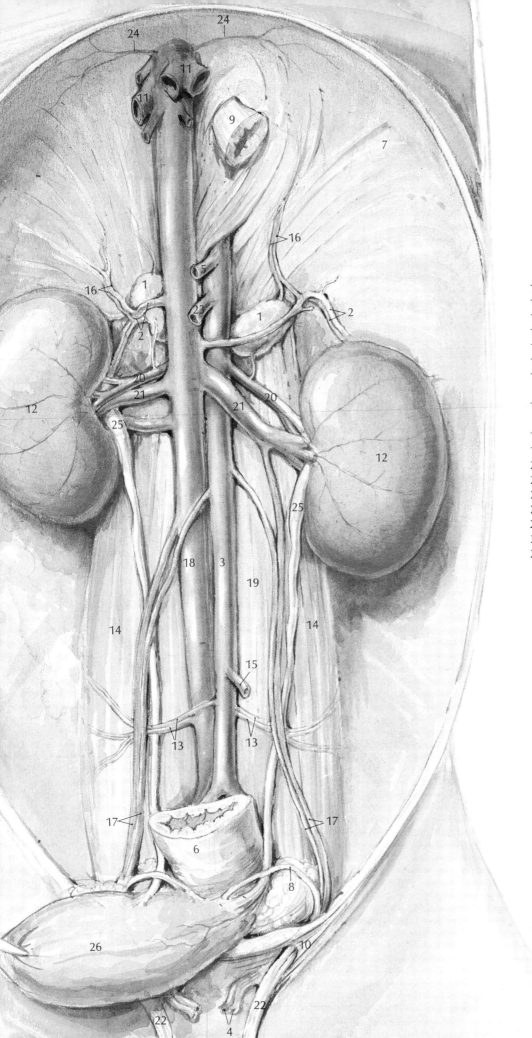

FIG. 39.
THE KIDNEYS AND
ABDOMINAL VESSELS
IN A MALE

1 adrenal gland
2 adrenolumbar artery
3 aorta
4 branch of deep femoral
    artery and vein
5 celiac artery
6 colon
7 diaphragm
8 ductus deferens
9 esophagus
10 external inguinal ring
11 hepatic veins
12 kidney
13 iliolumbar artery and vein
14 iliopsoas
15 inferior mesenteric artery
16 inferior phrenic artery and
    vein
17 internal spermatic artery
    and vein
18 postcava
19 psoas minor
20 renal artery
21 renal vein
22 spermatic cord
23 superior mesenteric artery
24 superior phrenic vein
25 ureter
26 urinary bladder

FIG. 40.
FRONTAL SECTION OF THE KIDNEY

1 arcuate artery and vein
2 cortex
3 dorsal branches of renal
  artery and vein
4 interlobar artery and vein
5 medulla
6 papilla
7 pelvis
8 ureter

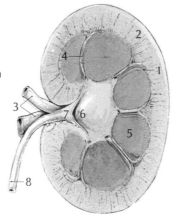

FIG. 41.
CROSS SECTION OF THE KIDNEY

1 cortex
2 dorsal branches of renal
  artery and vein
3 interlobar artery and vein
4 medulla
5 papilla
6 pelvis
7 ureter
8 ventral branches of renal
  artery and vein

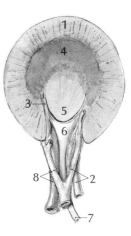

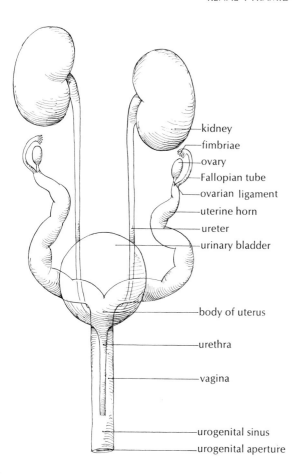

kidney
fimbriae
ovary
Fallopian tube
ovarian ligament
uterine horn
ureter
urinary bladder

body of uterus

urethra

vagina

urogenital sinus
urogenital aperture

NEPHRONS

Numerous small branches radiate from the arcuate vessels to supply the nephrons, microscopic tubules which constitute the functional units of the kidney. Each kidney contains several hundred thousand nephrons which follow an intricate course in the cortex and medulla. Urine, which is elaborated within the nephrons, drains into a series of collecting tubules which open onto the renal papilla.

RENAL PYRAMID

The term *renal pyramid* is applied to the papilla and the conical mass of collecting tubules which open onto it. In the cat there is but one renal pyramid. In man there are about twelve, and the human renal pelvis forms subdivisions termed calyces, each of which embraces one or two papillae.

Trace the ureter caudally. Near the bladder it passes dorsal to the ductus deferens, turns ventrally, and enters the bladder.

The urinary bladder is a musculomembranous sac; it is retroperitoneal and is attached to the abdominal walls by peritoneal folds termed ligaments. The medial ligament passes from the ventral side of the bladder to the linea alba. The two lateral ligaments, which contain sizable fat deposits, connect the sides of the bladder to the dorsal body wall. The free anterior portion of the bladder is the vertex; the attached posterior portion is the fundus. In the male the space between the bladder and the rectum is termed the rectovesical pouch; in the female the space between the bladder and the uterus is termed the vesicouterine pouch. Posteriorly the bladder is continuous with the urethra, via which urine passes to the exterior. Make a ventral incision in the bladder and open it to see the openings of the urethra and the ureters on the inner wall of the bladder.

The skin of the penis forms a sheath termed the prepuce. Pull it back and observe the glans, or enlarged distal end of the penis, which in the cat is covered with minute horny papillae. The scrotum is a pouch of skin which encloses the testes. Remove it and observe that each testis is enclosed in a fascial sac. (Do not open the fascial sac at this time.)

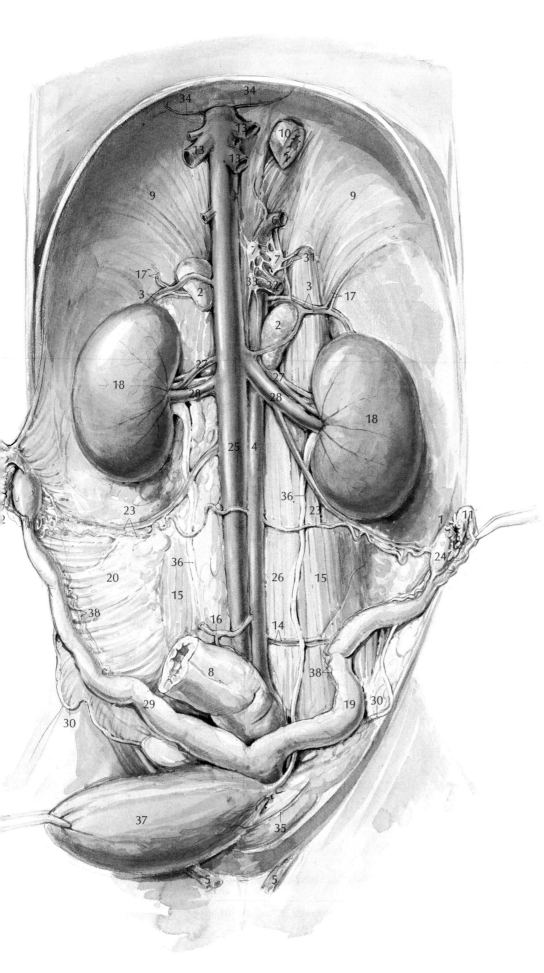

FIG. 42.
THE KIDNEYS, UTERUS,
AND ABDOMINAL VESSELS
IN A FEMALE

1 abdominal ostium of
   Fallopian tube
2 adrenal gland
3 adrenolumbar artery and
   vein
4 aorta
5 branches of deep femoral
   artery and vein
6 celiac artery
7 celiac ganglion
8 colon
9 diaphragm
10 esophagus
11 Fallopian tube
12 fimbriae
13 hepatic veins
14 iliolumbar artery and vein
15 iliopsoas
16 inferior mesenteric artery
17 inferior phrenic artery and
    vein
18 kidney
19 left uterine horn
20 mesometrium
21 mesovarium
22 mesosalpinx
23 ovarian artery and vein
24 ovary
25 postcava
26 psoas minor
27 renal artery
28 renal vein
29 right uterine horn
30 round ligament
31 splanchnic nerves
32 superior mesenteric
    artery
33 superior mesenteric
    ganglion
34 superior phrenic vein
35 umbilical artery
36 ureter
37 urinary bladder
38 uterine artery and vein

FIG. 43.
PELVIC PORTION OF THE MALE
UROGENITAL SYSTEM, LATERAL VIEW

1   anal sphincter
2   anus
3   aorta
4   bulbourethral gland
5   common iliac vein
6   crus of penis, cut
7   descending colon
8   ductus deferens
9   external iliac artery
10 fundus of bladder
11 glans
12 iliopsoas
13 penis
14 prostate gland
15 pubis
16 rectovesical pouch
17 rectum
18 right spermatic cord
19 sciatic nerve
20 testis within fascial sheath
21 thigh muscles, cut
22 ureter
23 urethra
24 urinary bladder
25 urogenital canal
26 urogenital opening
27 vertex of bladder

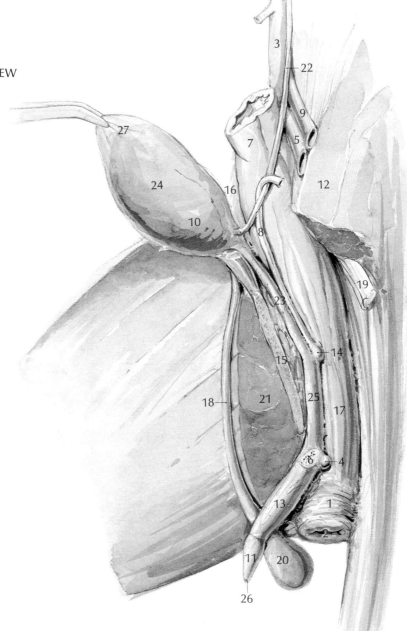

SPERMATIC CORD   The spermatic cord extends from the testis anteriorly to the abdominal wall. It is composed of the ductus deferens and the nerves, lymphatics, and vessels which supply the testis. Observe that the fascial sac of the testis continues as a thin covering around the components of the spermatic cord, and is continuous with the subcutaneous fascia and with the fascia of the external oblique muscle.

INGUINAL CANAL   The ductus deferens and the internal spermatic vessels pass through the abdominal wall via the inguinal canal. The external opening of the inguinal canal is termed the external inguinal ring; the internal opening of the inguinal canal is termed the internal inguinal ring. From the internal inguinal ring the ductus deferens turns toward the bladder, crossing over the ureter and the lateral ligament of the bladder.

DISSECTION OF PELVIS   The remaining components of the male reproductive system can best be seen in a lateral view of the pelvis. Remove the left leg near the hip joint. Remove and save the left testis within its fascial sac. Cut the pubic symphysis and cut the innominate bone above the acetabulum. Remove the cut portion of the innominate bone and clear away the pelvic muscles and connective tissue as necessary to identify the structures illustrated in Figure 43.

UROGENITAL CANAL

About the middle of the pubis the ductus deferens joins the urethra to form the urogenital canal, which extends to the urogenital opening at the end of the penis. Identify the prostate gland, near the union of the ductus deferens and the urethra. Paired bulbourethral glands open into the urogenital canal near the base of the penis.

CORPORA CAVERNOSA

Within the penis are two cylindrical structures termed corpora cavernosa; surrounding the urogenital canal is a similar, smaller structure termed the corpus spongiosum. These structures contain numerous blood sinuses which become distended during copulation. Near the pubic symphysis the two corpora cavernosa diverge, forming the right and left crura. Each crus is attached to the corresponding ramus of the ischium.

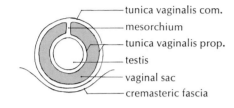

- tunica vaginalis com.
- mesorchium
- tunica vaginalis prop.
- testis
- vaginal sac
- cremasteric fascia

cross section of testis and tunics

Cut a thin cross section of the spermatic cord and examine it under a dissecting microscope. Observe that it consists of a tubular external fascial covering within which the ductus deferens and the internal spermatic vessels are suspended by a mesentery-like fold. Make a cross section through the middle of the left testis. Observe that the testis is attached to the surrounding fascial sac by a similar mesentery-like fold, the mesorchium.

DESCENT OF TESTES

The components of the fascial sac which encloses the testis are illustrated at the left in a schematic view of two successive stages in the descent of the testis. Each testis originates as a retroperitoneal structure near the kidneys. During fetal life the testis descends posteriorly through the inguinal canal and into the scrotum. The fascial sac, which encloses the testis, is formed by an outpouching of the abdominal wall, and each layer of the fascial sac is derived from a layer of the abdominal wall.

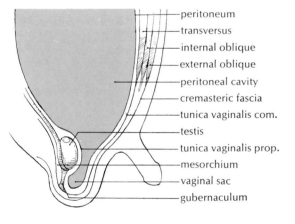

- peritoneum
- transversus
- internal oblique
- external oblique
- peritoneal cavity
- cremasteric fascia
- tunica vaginalis com.
- testis
- tunica vaginalis prop.
- mesorchium
- vaginal sac
- gubernaculum

The cremasteric fascia is derived from the fascia of the external oblique and from the subcutaneous fascia of the thigh. It covers only the ventral surface of the spermatic cord, whereas the tunica vaginalis propria and the tunica vaginalis communis completely surround the cord. The tunica vaginalis communis is continuous with the fascia of the transversus muscle. The tunica vaginalis propria is continuous with the peritoneum. It forms a double fold around the testis and the components of the spermatic cord. It consists of a parietal layer, closely adherent to the tunica vaginalis communis, and a visceral layer, closely adherent to the testis, ductus deferens, and spermatic vessels.

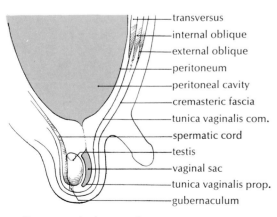

- transversus
- internal oblique
- external oblique
- peritoneum
- peritoneal cavity
- cremasteric fascia
- tunica vaginalis com.
- spermatic cord
- testis
- vaginal sac
- tunica vaginalis prop.
- gubernaculum

intermediate stages in descent of testes

Cut through the ventral side of the right fascial sac, exposing the testis. You have cut through the cremasteric fascia, the tunica vaginalis communis, and the parietal layer of the tunica vaginalis propria (these three layers are closely adherent and difficult to distinguish in gross dissection). The space surrounding the testis is the vaginal sac, an extension of the peritoneal cavity. The posterior end of the testis is attached to the fascial sac by a ligament, the gubernaculum, which is functional in the descent of the testes. It is homologous with the round ligament of the ovary.

Trim away the fascial sac from the testis and trace the ductus deferens toward the testis, noting its numerous convolutions. Near

FIG. 44.
PELVIC PORTION OF THE FEMALE
UROGENITAL SYSTEM, LATERAL VIEW

1 anal gland
2 anus
3 aorta
4 body of uterus
5 common iliac vein
6 colon
7 external iliac artery
8 iliolumbar artery and vein
9 iliopsoas
10 ilium
11 hypogastric artery
12 inferior gluteal artery
13 inferior mesenteric artery
14 left colic artery
15 left uterine horn
16 middle hemorrhoidal
     artery
17 postcava
18 pubis
19 rectum
20 rectus abdominis
21 sciatic nerve
22 superior gluteal artery
23 superior hemorrhoidal
     artery
24 thigh muscles, cut
25 umbilical artery
26 ureter
27 urethra
28 urinary bladder
29 urogenital aperture
30 urogenital sinus
31 uterine artery
32 vagina

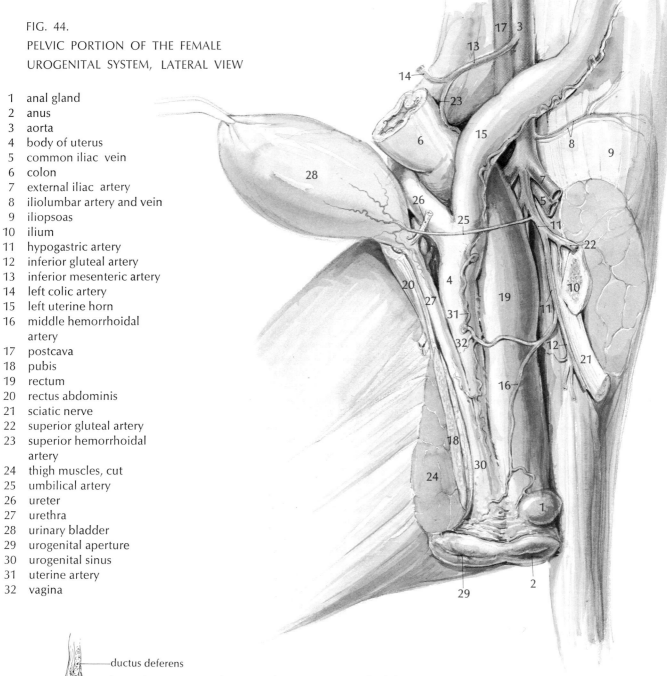

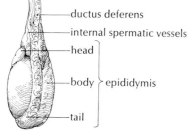

ductus deferens
internal spermatic vessels
head ⎤
body ⎬ epididymis
tail ⎦

right testis, medial view

OVARIES

the posterior end of the testis the ductus deferens is continuous
with the epididymis, which appears as a flattened band lying
along the dorsal side of the testis. The epididymis consists of an
anterior portion, the head, a middle portion, the body, and a
posterior portion, the tail. The head of the epididymis is connected
to the testis by numerous efferent ductules (too small to see in
gross dissection); they are continuous, via the convoluted ducts which
make up the epididymis, with the ductus deferens.

The ovaries lie just posterior to the kidneys. They are suspended
from the dorsal body wall by the mesovarium, a peritoneal fold
continuous with the broad ligament. Another ligament, the
ligament of the ovary, connects the ovary to the uterine horn.
The lumpy appearance of the ovary is due to the presence, within
the ovary, of Graafian follicles, each of which contains a developing
ovum. The Graafian follicles may best be seen by removing one
of the ovaries and cutting it in longitudinal section. Follicles which
have discharged their eggs are termed corpora lutea; these will

be seen best in pregnant animals.

FALLOPIAN TUBE   The Fallopian tube conveys the ovum from the ovary to the uterine horn. Its anterior end forms a funnel-shaped opening (the abdominal ostium) fringed by small, irregular projections termed fimbriae. Trace the Fallopian tube around the anterior end of the ovary and observe that it passes dorsal to the ovary to join the uterine horn. The mesentery of the Fallopian tube is the mesosalpinx. It is continuous with the broad ligament, and forms a pocket within which the ovary lies.

UTERUS   The uterus is a Y-shaped structure consisting of a central body and two uterine horns. The horns lie along either side of the dorsal wall of the abdominal cavity. The body of the uterus lies between the bladder and the rectum, and is continuous posteriorly with the vagina.

BROAD LIGAMENT   Each uterine horn is supported by a peritoneal fold termed the mesometrium. The mesometrium, mesosalpinx, and mesovarium are continuous with each other, and together they constitute the broad ligament. Also continuous with the broad ligament is the round ligament, a thin fibrous band which extends from the uterine horn to the abdominal wall. The anterior end of each broad ligament forms a free concave border near the ovary, and its attachment to the dorsal body wall extends in a curved line from a point near the posterior end of the kidneys to the lateral ligament of the bladder. From this point the broad ligament extends into the pelvis, serving to hold the body of the uterus and the vagina to the pelvic wall.

The remaining components of the female reproductive system can best be seen in a lateral view of the pelvis. Remove the left leg near the hip. Cut the pubic symphysis and cut the innominate bone above the acetabulum. Remove the cut portion of the innominate bone and clear away the pelvic muscles and connective tissue. Preserve the branches of the hypogastric artery as far as possible, and identify the structures illustrated in Figure 44.

Near the anterior border of the pubis the body of the uterus is continuous with the vagina. The opening between the body of the uterus and the vagina is the external uterine orifice. The posterior portion of the uterus forms a small rounded projection, termed the cervix, at the uterovaginal junction. The vagina extends from the cervix to the urethral orifice. The urethra lies along the ventral surface of the vagina, opening into it at the urethral orifice. The urogenital sinus is the common passage formed by the union of the urethra and the vagina. It opens at the urogenital aperture, just ventral to the anus. On either side of the urogenital aperture are folds of skin termed the labia majora. The urogenital aperture and labia together constitute the vulva. Near the urogenital aperture, in the ventral wall of the urogenital sinus, is the clitoris, homolog of the penis.

Remove the uterus, vagina, and urogenital sinus, and open them by trimming away the dorsal portion as illustrated in the marginal diagram.

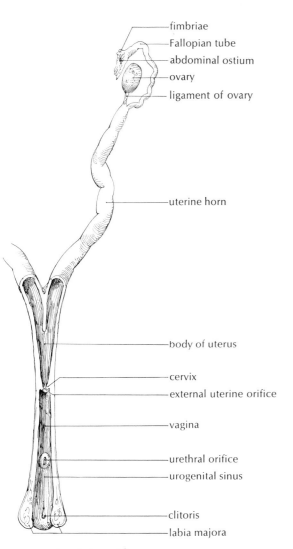

fimbriae
Fallopian tube
abdominal ostium
ovary
ligament of ovary

uterine horn

body of uterus

cervix
external uterine orifice

vagina

urethral orifice
urogenital sinus

clitoris
labia majora

dorsal view of the uterus

THE CIRCULATORY SYSTEM

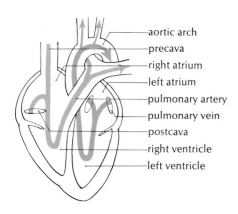

aortic arch
precava
right atrium
left atrium
pulmonary artery
pulmonary vein
postcava
right ventricle
left ventricle

Remove the pericardium and the thymus gland to expose the heart. Identify the vagus and phrenic nerves by referring to Figure 48, page 69, and Figure 69, page 95; be careful to preserve them.

The heart consists of two thin-walled atria and two muscular ventricles. Unoxygenated blood returns to the right atrium from the precava and postcava. From the right atrium it passes into the right ventricle, and from the right ventricle it is pumped through the pulmonary artery to the lungs. Oxygenated blood returns from the lungs via the pulmonary veins to the left atrium. From the left atrium it passes to the left ventricle, and from the left ventricle it is pumped to the body via the aorta. The ventricle and the atria are completely separated so that there is no mixture of oxygenated and unoxygenated blood within the heart.

PULMONARY VEINS

To identify the pulmonary veins pull the heart away from the lung on one side and examine the root of the lung. From each lobe a vein (usually uninjected) will be seen passing toward the dorsal side of the heart. There are three groups of pulmonary veins: the left pulmonary veins, from the anterior and middle lobes of the left lung; the right pulmonary veins, from the anterior and middle lobes of the right lung; and the dorsal pulmonary veins, from the terminal lobes of both lungs. Each group is composed of two or three veins (see Fig. 51, p. 72).

PULMONARY ARTERY

The pulmonary artery divides into right and left branches shortly after leaving the right ventricle. Near the point of division it is connected to the aorta by the ligamentum arteriosum, a strand of connective tissue representing the obliterated ductus arteriosus. (In the fetus the ductus arteriosus forms an open channel between the pulmonary artery and the aorta.) The left branch of the pulmonary artery passes ventral to the aorta to reach the left lung; the right branch passes between the aortic arch and the heart to reach the right lung.

Review the muscles of the upper forelimb and identify the muscles encountered in the dissection of the circulatory system as you come to them.

EXPOSURE OF BRACHIAL VESSELS

Cut and remove the pectoral muscles, making a dissection similar to Figure 45. Particular care should be taken not to damage the components of the brachial plexus, which are somewhat difficult to distinguish because they are surrounded by connective tissue and fat. In Figure 45 the musculocutaneous nerve has been omitted to expose the vessels lying below it. Identify the musculocutaneous nerve by referring to Figure 69 on page 95, and be careful to preserve it.

Individual variations in the nerves, arteries, and veins will be encountered in different specimens, and you should therefore examine several specimens other than your own in the course of this dissection.

Because the veins anterior to the heart lie for the most part ventral to the arteries, they will be described first. In your dissection, however, you should expose and identify the veins and arteries together. In most cases the veins follow correspondingly named arteries, and the veins will therefore not be described in detail.

The precava returns unoxygenated blood from the head and forelimbs to the right atrium. Refer to Figure 45 and identify its principal branches: the internal jugular, from the brain and spinal cord; the external jugular, from the head and neck; the subscapular, from the shoulder, and the axillary, from the forelimb.

Press the heart to the left and find the azygos vein, which arches over the root of the right lung and joins the precava near the right atrium. The azygos vein lies along the right side of the vertebral column in the thorax and receives the intercostal veins, as well as tributaries from the muscles of the dorsal abdominal wall, the esophagus, and the bronchi.

Toward the anterior end of the precava on the ventral side find the common stem by which the paired internal mammary veins enter the precava. The internal mammary veins lie on either side of the midline on the inner surface of the ventral chest wall, and anastomose with the superior epigastric veins.

The precava is formed by the union of the two short innominate veins. Dorsally each innominate vein receives the common stem of the vertebral and costocervical veins, corresponding in distribution to the arteries of the same name.

Near the first rib, the innominate vein is formed by the union of the external jugular and the subclavian veins. The subclavian vein is continuous with the axillary and brachial veins. The very short portion of this vein which lies within the thorax is the subclavian vein; the portion which lies within the axilla is the axillary vein; and the portion within the upper forelimb is the brachial vein. The largest tributary of the axillary vein is the subscapular, which joins it just lateral to the first rib. Near this point find the ventral thoracic vein, a small vessel joining the ventral surface of the axillary. Identify the long thoracic, thoracodorsal, and deep brachial veins as illustrated.

Near the elbow the median cubital vein forms a union between the cephalic vein and the brachial vein. Trace the course of the cephalic vein. It originates by superficial branches which unite near the wrist, and passes along the extensor side of the forelimb, dividing at the shoulder into two branches which anastomose with the posterior humeral circumflex and transverse scapular vein.

Near its union with the innominate vein the external jugular receives the internal jugular vein, which originates from the venous sinuses of the brain and from the vertebral column and the back

BRANCHES OF PRECAVA

transverse scapular

internal jugular
external jugular
vertebral
costocervical
subscapular
subclavian
axillary
innominate
ventral thoracic

internal mammary
precava

azygos

post. hum. circumflex

subscapular

subclavian

ventral thoracic

axillary

long thoracic

deep brachial

thoracodorsal

brachial

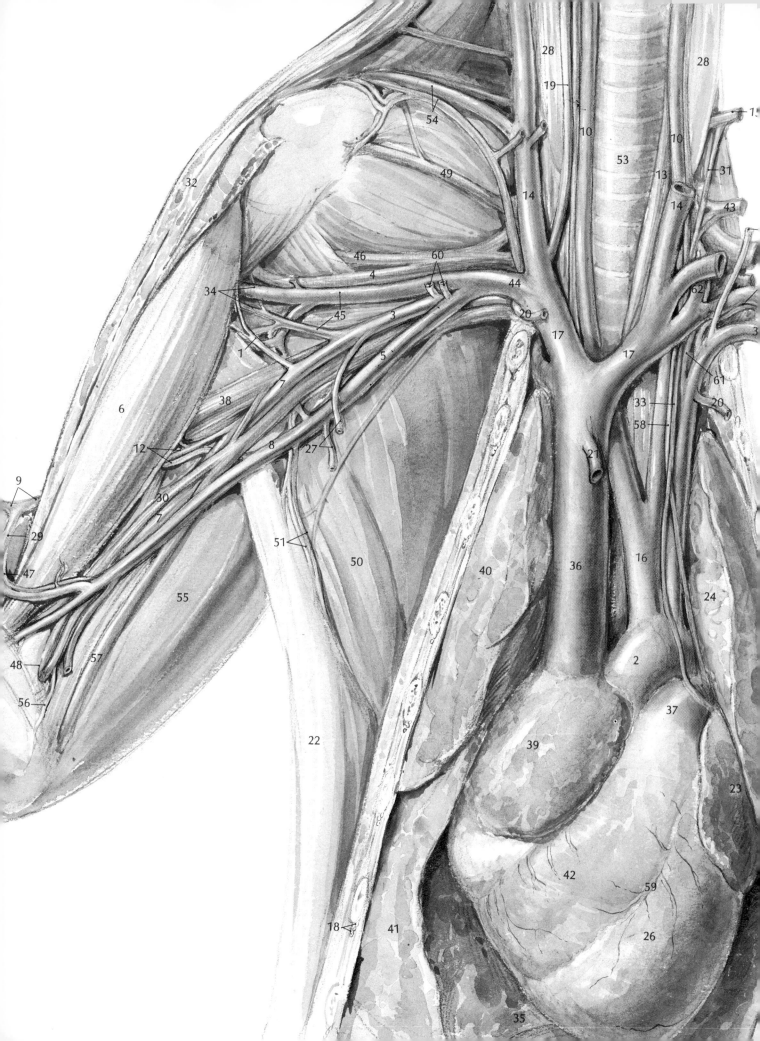

FIG. 45.

THE BRACHIAL VESSELS AND NERVES, VENTRAL VIEW

The musculocutaneous nerve is removed to expose underlying structures.
For details of the brachial plexus, see Figure 69, page 95.

1 anterior humeral circumflex artery	22 latissimus dorsi	45 subscapular artery and vein
2 aortic arch	23 left atrium	46 subscapular nerve
3 axillary artery	24 left lung, anterior lobe	47 superior radial collateral artery
4 axillary nerve	25 left lung, middle lobe	48 supracondyloid foramen
5 axillary vein	26 left ventricle	49 suprascapular nerve
6 biceps	27 long thoracic artery and vein	50 teres major
7 brachial artery	28 longus capitis	51 thoracodorsal artery and vein
8 brachial vein	29 median cubital vein	52 thyrocervical artery
9 cephalic vein	30 median nerve	53 trachea
10 common carotid artery	31 origin of phrenic nerve	54 transverse scapular artery and vein
11 costocervical artery and vein	32 pectoral muscles, cut	55 triceps
12 deep brachial artery and vein	33 phrenic nerve	56 ulnar collateral artery
13 esophagus	34 posterior humeral circumflex artery and vein	57 ulnar nerve
14 external jugular vein	35 postcava	58 vagus nerve
15 fifth cervical nerve	36 precava	59 ventral branches of left coronary artery and vein
16 innominate artery- Brachiocephilic artery	37 pulmonary artery	60 ventral thoracic artery and vein
17 innominate vein- Brachiocephilic veins	38 radial nerve	61 vertebral artery
18 intercostal vein, artery, and nerve	39 right atrium	62 vertebral vein
19 internal jugular vein	40 right lung, anterior lobe	
20 internal mammary artery	41 right lung, middle lobe	
21 internal mammary vein	42 right ventricle	
	43 sixth cervical nerve	
	44 subclavian vein	

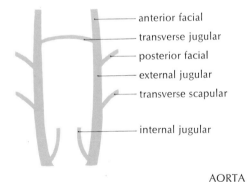

— anterior facial
— transverse jugular
— posterior facial
— external jugular
— transverse scapular
— internal jugular

of the head.

The external jugular vein, larger and more superficial than the internal jugular vein, is formed by the union of the anterior and posterior facial veins (see Figs. 15 and 16 on pp. 20 and 21). Near the jaw the external jugular veins are connected by the transverse jugular vein. At the shoulder the external jugular vein receives the large transverse scapular vein and one or more small tributaries from nearby muscles.

Remove the precava and its branches to make a dissection similar to Figure 46.

AORTA

The aorta is the large arterial trunk which conveys oxygenated blood from the left ventricle to the body. At its origin it makes an abrupt curve to the left, passing dorsal to the pulmonary artery and continuing down the left side of the vertebral column to the pelvis, where it divides into branches that supply the legs. The portion of the aorta anterior to the diaphragm is termed the thoracic aorta, and the proximal curved portion of the thoracic aorta is termed the aortic arch.

CORONARY ARTERIES

The coronary arteries arise near the origin of the aorta, just cranial to the aortic valves (see Fig. 53, p. 74). The left coronary artery passes dorsal to the pulmonary artery and divides into two branches: one to the dorsal and one to the left and ventral sides of the heart. The right coronary artery passes to the right, lying between the right auricle and the right ventricle, giving branches to

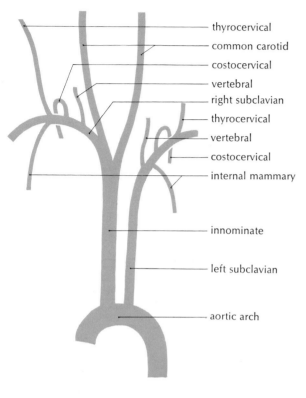

— thyrocervical
— common carotid
— costocervical
— vertebral
— right subclavian
— thyrocervical
— vertebral
— costocervical
— internal mammary
— innominate
— left subclavian
— aortic arch

BRANCHES OF AXILLARY ARTERY

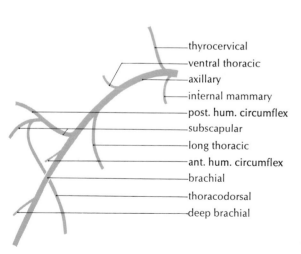

— thyrocervical
— ventral thoracic
— axillary
— internal mammary
— post. hum. circumflex
— subscapular
— long thoracic
— ant. hum. circumflex
— brachial
— thoracodorsal
— deep brachial

the right side of the heart.

The aortic arch gives rise to the innominate and the left subclavian arteries, which supply the head and forelimbs. About the level of the second rib the innominate divides into right subclavian and right and left common carotids. Near the first rib the subclavian gives off four branches: the internal mammary, vertebral, costocervical, and thyrocervical.

The internal mammary arises from the ventral surface of the subclavian and passes caudally in the ventral thoracic wall, giving off branches to the adjacent muscles, the pericardium, the mediastinum, and the diaphragm. It passes out of the thorax and anastomoses with the inferior epigastric artery.

The vertebral artery arises from the dorsal surface of the subclavian and passes cranially through the vertebrarterial canal of the cervical vertebrae, giving off branches to the deep neck muscles and to the spinal cord. Near the foramen magnum the right and left vertebral arteries enter the vertebral canal and unite to form the basilar artery, which lies along the ventral aspect of the medulla oblongata.

The costocervical artery arises from the dorsal surface of the subclavian. It sends branches to the deep muscles of the neck and shoulder and to the first two costal interspaces.

The thyrocervical artery arises from the cranial aspect of the subclavian and passes cranially and laterally, supplying the muscles of the neck and chest. At the cranial border of the scapula it is termed the transverse scapular artery, and divides into several branches which supply the shoulder muscles. The branches of the transverse scapular artery accompany the fifth cervical and the suprascapular nerves.

Lateral to the first rib the subclavian artery continues into the axilla as the axillary artery.

The ventral thoracic artery arises from the ventral surface of the axillary just lateral to the first rib, passing caudally to supply the medial ends of the pectoral muscles. It accompanies the anterior ventral thoracic nerve.

The long thoracic artery arises lateral to the ventral thoracic, passing caudally to the pectoral muscles and the latissimus dorsi. It accompanies the posterior ventral thoracic nerve.

The subscapular artery is the largest branch of the axillary. It passes laterally and dorsally between the long head of the triceps and the latissimus dorsi to supply the dorsal shoulder muscles. A short distance from its origin the subscapular gives off two branches, the thoracodorsal and the posterior humeral circumflex. The thoracodorsal artery passes dorsal to the brachial plexus, giving branches to the teres major and the latissimus dorsi. The posterior humeral circumflex passes dorsally with the axillary nerve between the long and lateral heads of the triceps. To trace the branches of the posterior humeral circumflex and the subscapular arteries, turn the specimen over and remove the spinodeltoid muscle. Branches of these arteries will then be seen near the origin of the triceps.

Distal to the origin of the subscapular artery, the axillary continues as the brachial artery. The anterior humeral circumflex arises just distal to the origin of the subscapular and supplies the biceps. The deep brachial accompanies the radial nerve to the dorsal side of the

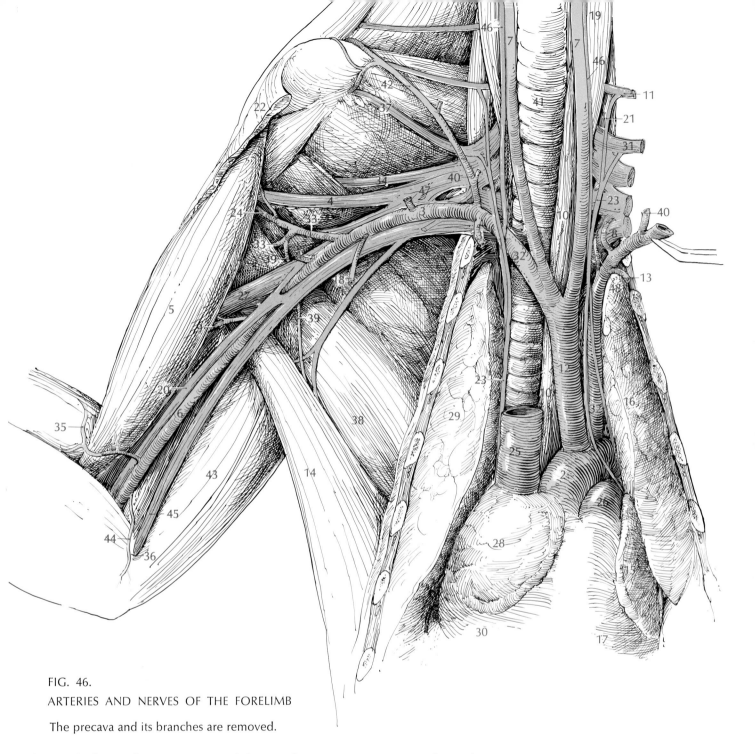

FIG. 46.

ARTERIES AND NERVES OF THE FORELIMB

The precava and its branches are removed.

1	anterior humeral circumflex artery	17	left ventricle	33	subscapular artery
2	aortic arch	18	long thoracic artery	34	subscapular nerve
3	axillary artery	19	longus capitis	35	superior radial collateral artery
4	axillary nerve	20	median nerve	36	supracondyloid foramen
5	biceps	21	origin of phrenic nerve	37	suprascapular nerve
6	brachial artery	22	pectoral muscles, cut	38	teres major
7	common carotid artery	23	phrenic nerve	39	thoracodorsal artery
8	costocervical artery	24	posterior humeral circumflex artery	40	thyrocervical artery
9	deep brachial artery	25	precava	41	trachea
10	esophagus	26	pulmonary artery	42	transverse scapular artery
11	fifth cervical nerve	27	radial nerve	43	triceps
12	innominate artery	28	right atrium	44	ulnar collateral artery
13	internal mammary artery	29	right lung, anterior lobe	45	ulnar nerve
14	latissimus dorsi	30	right ventricle	46	vagus nerve
15	left atrium	31	sixth cervical nerve	47	ventral thoracic artery
16	left lung, anterior lobe	32	subclavian artery	48	vertebral artery

forelimb. Distal to the deep brachial the brachial artery gives off several muscular branches, and just proximal to the elbow it gives off two collateral branches. It then passes through the supracondyloid foramen of the humerus in company with the median nerve. Distal to the elbow the brachial artery is continuous with the radial, and gives rise to the ulnar artery.

To identify the branches of the right common carotid artery remove the digastric, and cut and pull back the sternomastoid and cleidomastoid muscles to make a dissection similar to Figure 47. In tracing these arteries be careful to avoid injury to the cervical nerves.

The superior thyroid artery branches from the common carotid at the cranial end of the thyroid gland. It supplies the thyroid gland, superficial laryngeal muscles, and ventral neck muscles. At the same level the common carotid gives off one or more dorsal muscular branches to the deep muscles of the neck.

Near the point where it is crossed by the hypoglossal nerve the common carotid gives off the occipital and the internal carotid arteries, which sometimes arise by a common stem. The occipital supplies the deep muscles of the neck and extends toward the dorsal side of the tympanic bulla, continuing to the back of the skull where it runs along the superior nuchal line. The internal carotid artery passes toward the ventral side of the tympanic bulla and enters the bulla with the Eustachian tube; it then enters the skull via the foramen lacerum and joins the posterior cerebral artery.

After giving off the internal carotid artery, the common carotid continues as the external carotid artery. It passes deep to the digastric and gives off the lingual artery, which accompanies the hypoglossal nerve, giving off branches to the hyoid and pharyngeal muscles. It then enters the tongue, to which it gives numerous branches.

The external maxillary artery branches from the external carotid near the angle of the jaw. It passes along the ventral border of the masseter muscle, extending toward the nose and giving off branches to the lips and mouth. Distal to the external maxillary, the external carotid gives off the posterior auricular and superficial temporal branches, which run dorsally and laterally, supplying the superficial muscles of the side and back of the head.

After giving off the superficial temporal, the external carotid turns medially near the posterior margin of the masseter and continues as the internal maxillary. After giving off inferior alveolar and middle meningeal branches, it ramifies to form the carotid plexus, a network of arteries which surrounds the maxillary branch of the fifth cranial nerve near the foramen rotundum. The internal maxillary and the carotid plexus give various branches to the brain, eye, and other deep structures of the head. See the carotid plexus as illustrated in Figure 73, page 103.

Remove the left lung and cut away the left lateral portion of the thorax. Remove connective tissue and pleura as necessary to make a dissection similar to Figure 48. Near the aorta on the left side you will find the thoracic duct, which appears as a thin tube of irregular diameter. It may be removed to expose the underlying structures. (The lymphatics will be seen best in a demonstration dissection of an animal in which they have been injected).

The thoracic aorta gives off paired intercostal arteries corresponding to the interspaces between the last eleven ribs. The intercostal

EXPOSURE OF COMMON CARTOTID ARTERY

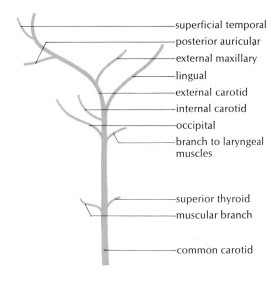

- superficial temporal
- posterior auricular
- external maxillary
- lingual
- external carotid
- internal carotid
- occipital
- branch to laryngeal muscles

- superior thyroid
- muscular branch

- common carotid

DISSECTION OF THORAX

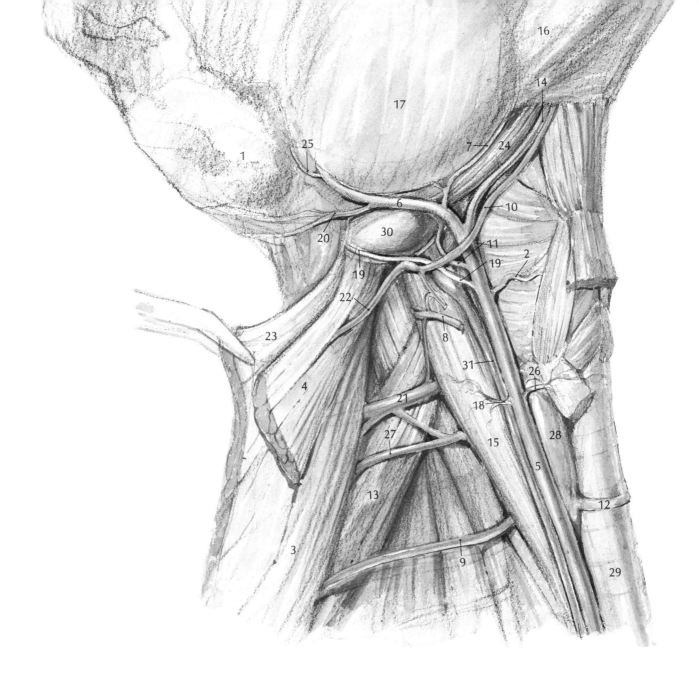

FIG. 47.

THE RIGHT COMMON CAROTID ARTERY

1 auricular cartilage	18 muscular branch
2 branch to laryngeal muscles	19 occipital artery
3 clavotrapezius	20 posterior auricular artery
4 cleidomastoid	21 second cervical nerve
5 common carotid artery	22 spinal accessory nerve (11)
6 external carotid artery	23 sternomastoid
7 external maxillary artery	24 styloglossus
8 first cervical nerve	25 superficial temporal artery
9 fourth cervical nerve	26 superior thyroid artery
10 hypoglossal nerve (12)	27 third cervical nerve
11 internal carotid artery	28 thyroid gland
12 isthmus of thyroid gland	29 trachea
13 levator scapulae ventralis	30 tympanic bulla
14 lingual artery	31 vagus nerve (10) and sympathetic trunk
15 longus capitis	
16 mandible	
17 masseter	

arteries corresponding to the first and second costal interspaces arise as branches of the costocervical artery.

The thoracic aorta also gives off paired bronchial arteries to the bronchi and several esophageal arteries of varying origin.

The aorta passes through the diaphragm at the level of the second lumbar vertebra. The portion of the aorta between the diaphragm and the pelvis is termed the abdominal aorta.

The first two branches given off by the abdominal aorta are the celiac and the superior mesenteric. The branches of these arteries were identified during the dissection of the alimentary canal (pp. 47-48). Other branches of the abdominal aorta are illustrated in Figures 39 and 42, pages 53 and 55.

Just below the superior mesenteric artery, the abdominal aorta gives off the paired adrenolumbar arteries, which pass laterally along the dorsal body wall. Each adrenolumbar artery gives off an inferior phrenic artery to the diaphragm and an adrenal artery to the adrenal gland. It then continues laterally to supply the muscles of the dorsal body wall.

The paired renal arteries supply the kidneys and, in some specimens, give rise to the adrenal artery (Figure 39, page 53).

Before entering the substance of the kidney each renal artery usually divides into two or more branches.

The paired internal spermatic arteries (in the male) arise from the aorta near the caudal ends of the kidneys. They lie on the surface of the psoas minor and iliopsoas muscles, passing caudally to the internal inguinal ring, and through the inguinal canal to accompany the ductus deferens to the testes.

The paired ovarian arteries (in the female) arise from the aorta near the caudal ends of the kidneys, passing laterally in the broad ligament to supply the ovaries. Each ovarian artery gives a branch to the cranial end of the corresponding uterine horn. This branch anastomoses with the uterine artery, a branch of the middle hemorrhoidal.

Seven pairs of lumbar arteries arise from the dorsal surface of the aorta, supplying the muscles of the dorsal abdominal wall.

At the level of the last lumbar vertebra the inferior mesenteric artery arises from the aorta. Near its origin it divides into the left colic artery, which passes anteriorly to supply the descending colon, and the superior hemorrhoidal, which passes posteriorly to supply the rectum.

The paired iliolumbar arteries arise near the inferior mesenteric and pass laterally across the psoas minor and the iliopsoas muscles to supply the muscles of the dorsal abdominal wall.

POSTCAVA

The postcava returns venous blood from the hindlimbs and abdomen to the right atrium. It receives iliolumbar veins, lumbar veins, ovarian or internal spermatic veins, renal veins, and adrenolumbar veins corresponding to the arteries of the same name. (The left internal spermatic or ovarian vein is usually a tributary of the left renal vein.) The postcava does not, however, receive tributaries corresponding to the celiac, superior mesenteric, or inferior mesenteric arteries. Blood from these arteries returns via the portal vein, passes through the sinusoids of the liver, and enters the postcava via the hepatic veins. The postcava passes through the liver, within which it receives the hepatic veins, and pierces the diaphragm near

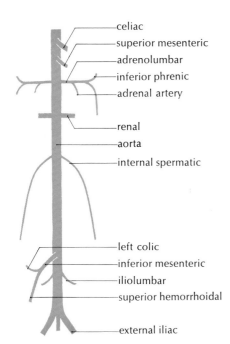

celiac
superior mesenteric
adrenolumbar
inferior phrenic
adrenal artery
renal
aorta
internal spermatic
left colic
inferior mesenteric
iliolumbar
superior hemorrhoidal
external iliac

FIG. 48.

THE HEART AND THORACIC AORTA, LATERAL VIEW

1 adrenal gland
2 aortic arch
3 branches of left bronchus
4 common carotid artery
5 celiac artery
6 celiac ganglion
7 coronary artery and vein
8 diaphragm
9 dorsal branch of left
   vagus nerve
10 dorsal division of
   vagus nerve
11 esophagus
12 eighth cervical nerve
13 first thoracic nerve
14 greater splanchnic nerve
15 inferior cervical ganglion
16 innominate artery
17 intercostal artery
18 intercostal vein, artery,
   and nerve
19 internal mammary artery
20 kidney
21 left atrium
22 left branch of
   pulmonary artery
23 left subclavian artery
24 left ventricle
25 lesser splanchnic nerve
26 phrenic nerve
27 precava
28 pulmonary artery
29 pulmonary vein
30 recurrent laryngeal nerve
31 right auricle
32 right lung,
   anterior lobe
33 right ventricle
34 sixth cervical nerve
35 seventh cervical nerve
36 subclavian artery
37 superior mesenteric artery
38 superior mesenteric
   ganglion
39 sympathetic trunk
40 thoracic aorta
41 trachea
42 vagus nerve
43 ventral branch of
   right vagus nerve
44 ventral division of
   vagus nerve

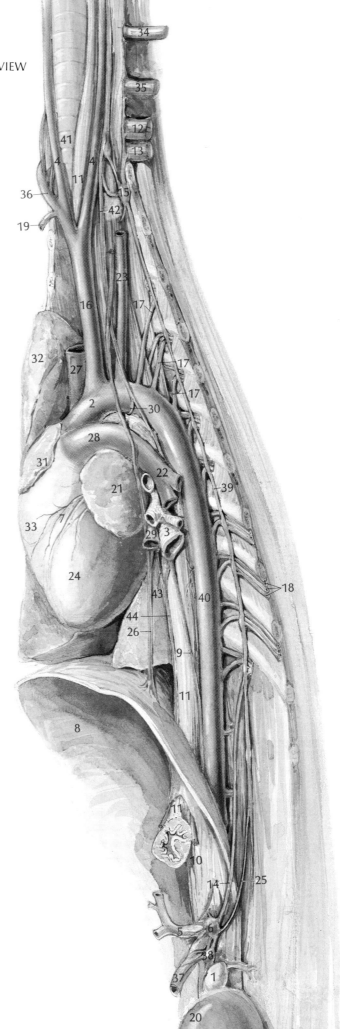

FIG. 49.
PELVIC VESSELS AND NERVES, LATERAL VIEW

1 aorta
2 caudal vertebra
3 common iliac v.
4 deep femoral a. & v.
5 external iliac a.
6 external iliac v.
7 flexor of tail
8 hypogastric a.
9 hypogastric v.
10 iliolumbar a. & v.
11 iliopsoas m.
12 inferior epigastric a. & v.
13 inferior gluteal a. & v.
14 levator ani m.
15 lumbar n. #8
16 median sacral a. & v.
17 middle hemorrhoidal a. & v.
18 obturator nerve
19 postcava
20 psoas minor m.
21 pubic symphysis
22 rectus abdominis m.
23 sacral nerve #1
24 sacrum
25 superior gluteal a.
26 umbilical a.

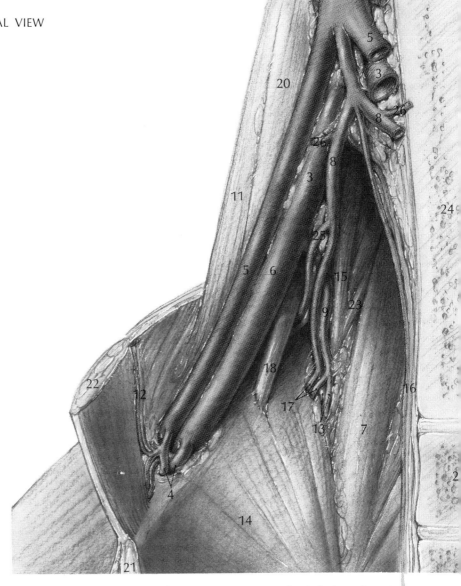

EXPOSURE OF PELVIC VESSELS

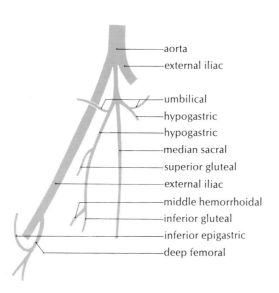

the central tendon. At this point it receives the inferior phrenic veins from the diaphragm. It then passes between the heart and the caudal lobe of the right lung to enter the right atrium.

Remove the pelvic viscera and identify the vessels illustrated in Figure 49. Also see the blood supply of the pelvic viscera as illustrated in Figure 44 on page 58.

Near the sacrum the aorta divides into right and left external iliac arteries. From this point of division a short common stem extends caudally, giving rise to the paired hypogastric arteries and the unpaired median sacral artery, which passes along the ventral aspect of the tail.

The umbilical artery is the first branch of the hypogastric. It arises near the origin of the hypogastric and passes ventrolaterally within the fat surrounding the lateral ligament of the bladder, dividing into numerous branches on the surface of the bladder.

The superior gluteal artery is a large branch of the hypogastric which passes dorsolaterally, supplying the gluteal muscles and other muscles of the thigh. The middle hemorrhoidal passes first ventrally and then caudally along the lateral surface of the rectum, extending as far as the anus and supplying adjacent structures. Near its origin it gives off a branch which passes ventrally to the urethra.

FIG. 50.

VESSELS OF THE PELVIS AND THIGH, VENTRAL VIEW

1 adductor femoris
2 adductor longus
3 aorta
4 common iliac vein
5 deep femoral artery
  and vein
6 external iliac artery
7 external iliac vein
8 femoral artery
9 femoral nerve
10 femoral vein
11 genitofemoral nerve
12 gracilis
13 hypogastric artery
14 hypogastric vein
15 iliolumbar artery and vein
16 iliopsoas
17 inferior epigastric artery
  and vein
18 lateral femoral circumflex
  artery and vein
19 medial branch of third
  lumbar nerve
20 median sacral artery
  and vein
21 muscular branches of
  femoral artery and vein
22 obturator nerve
23 postcava
24 psoas minor
25 rectus femoris
26 saphenous nerve
27 saphenous nerve,
  artery, and vein
28 sartorius
29 superior articular
  artery and vein
30 superior gluteal artery
31 umbilical artery
32 vastus medialis

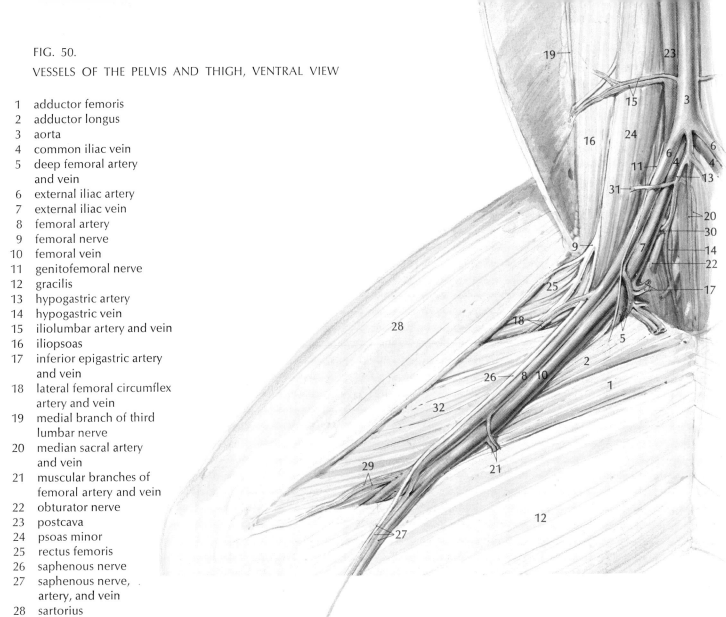

This branch is small in the male; in the female it is the prominent uterine artery, which passes anteriorly along the body of the uterus and the uterine horn, extending to the anterior end of the horn and anastomosing with a branch of the ovarian artery. The terminal branch of the hypogastric is the inferior gluteal artery. It follows the sciatic nerve, dividing into several branches.

Trim away the medial border of the sartorius muscle and clear away fat and connective tissue as necessary to identify the structures illustrated in Figure 50.

Just before leaving the abdominal cavity the external iliac artery gives off the deep femoral artery, which passes between the iliopsoas and the pectineus to supply the muscles of the thigh. Near its origin it gives off several branches. One of these is the inferior epigastric artery, which passes anteriorly on the inner surface of the rectus abdominis muscle, and anastomoses with the terminal branches of the internal mammary artery. One branch of the deep femoral passes to the bladder, and another pierces the abdominal wall to ramify on the medial aspect of the thigh, contributing a small vessel (the external spermatic artery) to the spermatic cord.

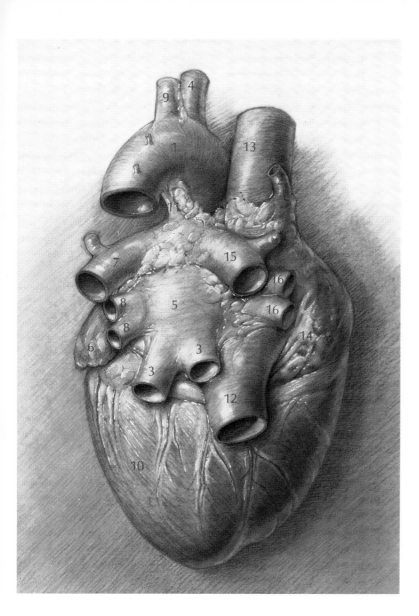

FIG. 51.
DORSAL VIEW OF THE
HEART OF THE CAT

 1   aorta
 2   azygos vein
 3   dorsal pulmonary vein
 4   innominate artery
 5   left atrium
 6   left auricle
 7   left pulmonary artery
 8   left pulmonary veins
 9   left subclavian artery
10   left ventricle
11   ligamentum arteriosum
12   postcava
13   precava
14   right atrium
15   right pulmonary artery
16   right pulmonary veins
17   right ventricle

The external iliac artery continues outside the abdominal cavity as the femoral artery, lying on the medial surface of the thigh. Its first branch is the lateral femoral circumflex artery, which passes laterally between the rectus femoris and the vastus medialis. About the middle of the thigh a large muscular branch is given off medially. Near the knee the superior articular artery is given off laterally; it passes toward the knee. At about the same point the saphenous artery is given off medially. It follows a superficial course along the medial surface of the lower limb as far as the foot. The femoral artery then passes between the vastus medialis and the semimembranosus. It continues as the popliteal artery (in the region of the knee) and then branches into anterior and posterior tibial arteries to supply the lower limb and foot. The veins of the pelvis and thigh correspond to the arteries, except that the hypogastric vein joins the external iliac vein directly to form the common iliac vein; notice that there is no common iliac artery in the cat.

Remove the heart, leaving as much as possible of the pulmonary vessels attached to the heart. Remove connective tissue and fat as necessary to identify the vessels illustrated in Figure 51.

DISSECTION OF RIGHT ATRIUM

Trim away the lateral wall of the right atrium and remove any coagulated blood and latex found within. Probe the precava

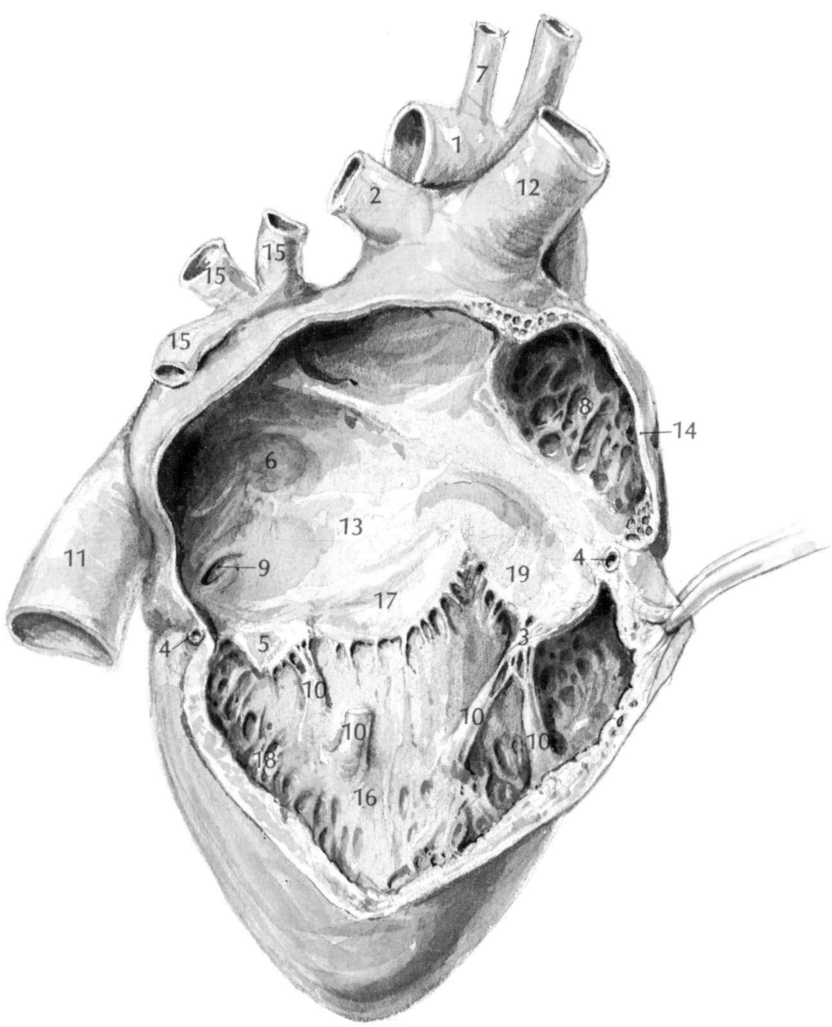

FIG. 52.
THE RIGHT ATRIUM AND RIGHT VENTRICLE

1   aorta
2   azygos vein
3   chordae tendineae
4   coronary artery
5   dorsal cusp of
    tricuspid valve
6   fossa ovalis
7   left subclavian artery
8   musculi pectinati
9   opening of coronary sinus
10 papillary muscles
11 postcava
12 precava
13 right atrium
14 right auricle
15 right pulmonary veins
16 right ventricle
17 septal cusp of tricuspid
    valve
18 trabeculae carneae
19 ventral cusp of tricuspid
    valve

and postcava, observing that they both empty into the right atrium.
In the dorsal atrial wall, just caudal to the opening of the postcava,
find a small oblong opening guarded by a valve. This is the opening
of the coronary sinus, via which blood returns from the coronary
veins to the right atrium. Probe the opening, observing that it
leads to a venous channel on the dorsal side of the heart. This
channel is the coronary sinus; most of the veins of the heart empty
into it.

The term *auricle* is applied to the portion of the atrial wall
that forms a flap-like extension projecting ventrally over the coronary
vessels. Observe the delicate strands which form a network on
its inner surface; they are termed the musculi pectinati. Near the
opening of the postcava find the fossa ovalis, a shallow oval
depression in the medial wall of the atrium. This is the site of the
foramen ovale in fetal life. Soon after birth the foramen ovale is
closed by a membrane, which completely separates the right and
left atria. The right atrioventricular opening is guarded by the

DISSECTION OF RIGHT VENTRICLE

tricuspid valve. Observe its cranial surface; then cut open the right
ventricle and trim away portions of the ventricular wall to make
a dissection similar to Figure 52. The tricuspid valve consists of three
cusps named according to their positions: dorsal, ventral, and septal.
Slender strands termed chordae tendineae are connected to the

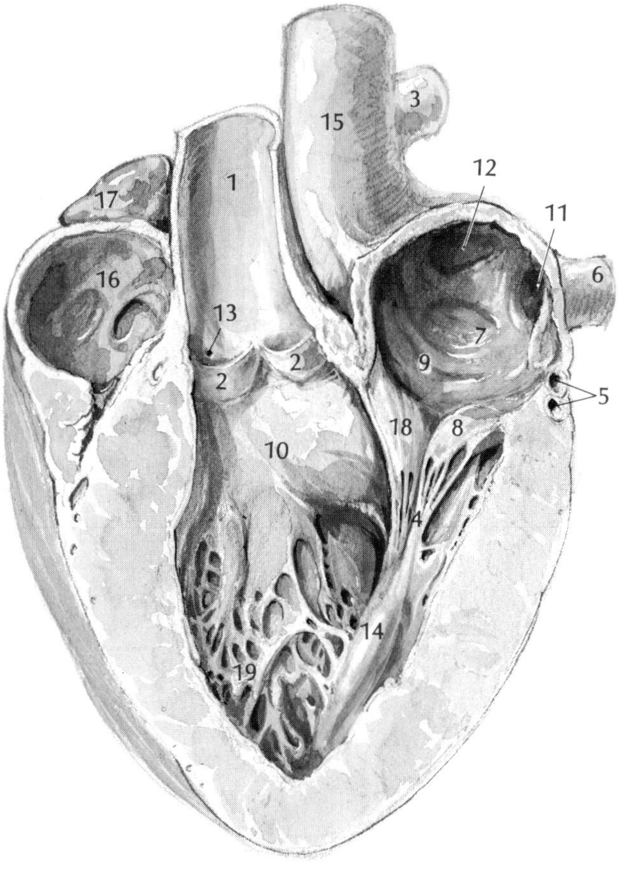

FIG. 53.
THE LEFT ATRIUM AND LEFT VENTRICLE

1　aorta
2　aortic semilunar valves
3　azygos vein
4　chordae tendineae
5　coronary artery and vein
6　dorsal pulmonary vein
7　fossa ovalis
8　lateral cusp of bicuspid
　　valve
9　left atrium
10　left ventricle
11　opening of dorsal
　　pulmonary vein
12　opening of right
　　pulmonary vein
13　origin of right
　　coronary artery
14　papillary muscle
15　precava
16　right atrium
17　right auricle
18　septal cusp of
　　bicuspid valve
19　trabeculae carneae

DISSECTION OF LEFT ATRIUM AND VENTRICLE

free margins of the cusps. These strands connect the septal cusp directly to the ventricular septum, and attach the dorsal and ventral cusps to muscular projections termed papillary muscles. The irregular muscular strands on the inner wall of the ventricle are termed trabeculae carneae.

Make a slit in the ventral wall of the right ventricle and the pulmonary artery. Observe the three pulmonary semilunar valves at the base of the pulmonary artery.

Referring to Figure 53, remove the pulmonary artery and trim away the lateral walls of the left atrium and ventricle. Cut the medial wall of the left atrium and make a slit in the left lateral wall of the aorta, observing the three aortic semilunar valves at the base of the aorta. Near two of the semilunar valves find the openings of the coronary arteries. The right and left atria are completely separated by the interatrial septum. Similarly, the right and left ventricles are completely separated by the interventricular septum. If the interatrial septum is observed by transmitted light, it will be seen that the fossa ovalis is the thinnest part of the septum.

Trace the pulmonary veins to their openings into the left atrium. The left atrioventricular opening is guarded by the bicuspid valve, consisting of two cusps, septal and lateral. Both cusps are attached to the ventricular wall by chordae tendineae and papillary muscles. Note that the wall of the left ventricle is considerably thicker than the wall of the right ventricle.

THE NERVOUS SYSTEM

CENTRAL NERVOUS SYSTEM

PERIPHERAL NERVOUS SYSTEM

CEREBROSPINAL NERVES

AUTONOMIC SYSTEM
SYMPATHETIC SYSTEM

PARASYMPATHETIC SYSTEM

The nervous system consists of two parts: the central nervous system and the peripheral nervous system. The central nervous system includes the brain and spinal cord. The peripheral nervous system consists of the nerves which connect the central nervous system with the various parts of the body. The peripheral nervous system is made up of two groups of nerves: cerebrospinal and autonomic.

The cerebrospinal nerves include the twelve pairs of cranial nerves, which are attached to the base of the brain, and the thirty-eight pairs of spinal nerves, which are attached to the spinal cord.

The autonomic system consists of the sympathetic and the parasympathetic systems. The portions of the sympathetic system which are grossly visible in the cat are the sympathetic trunks, strands of nerve tissue which lie on either side of the vertebral column from the base of the skull to the tail, and certain ganglia and nerves connected to these trunks. The parasympathetic system consists of components of cranial nerves 3, 7, 9, and 10, and of certain branches which accompany the sacral nerves. The components of the parasympathetic system will not be distinguished grossly in the cat.

FIG. 54.
THE THREE PRIMARY DIVISIONS
OF THE BRAIN

1   mesencephalon
2   prosencephalon
3   rhombencephalon

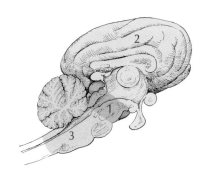 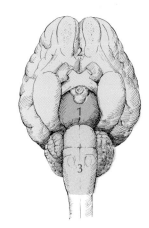

The brain consists of three primary divisions: the rhombencephalon, the mesencephalon, and the prosencephalon. See a demonstration dissection in which these divisions are separated from each other as shown in Figure 54.

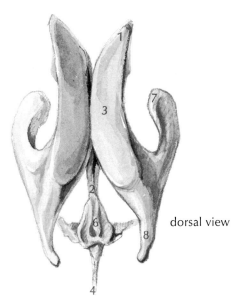

dorsal view

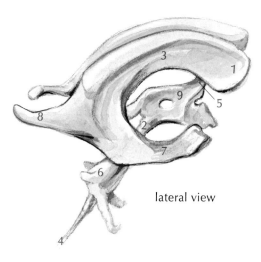

lateral view

FIG. 55.
THE VENTRICLES OF THE
HUMAN BRAIN (after Anson)

1 anterior horn of
   lateral ventricle
2 aqueduct
3 body of lateral ventricle
4 central canal of spinal
   cord
5 foramen of Monro
6 fourth ventricle
7 inferior horn of lateral
   ventricle
8 posterior horn of lateral
   ventricle
9 third ventricle

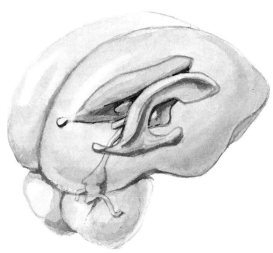

VENTRICLES   Within the brain are four intercommunicating spaces termed ventricles. The lateral (first and second) ventricles are paired cavities which lie within the prosencephalon on either side of the midline. Each lateral ventricle communicates, via a channel termed the foramen of Monro, with the third ventricle, which lies within the prosencephalon in the median sagittal plane. Posteriorly the third ventricle is continuous with a narrow channel termed the aqueduct, which lies within the mesencephalon. Within the rhombencephalon the aqueduct widens to form the fourth ventricle. Posteriorly the fourth ventricle is continuous with the central canal of the spinal cord.

PROSENCEPHALON   The prosencephalon includes the telencephalon and the diencephalon. The diencephalon consists of the third ventricle and the structures immediately surrounding it. The telencephalon may be thought of as identical with the cerebrum. (The optic chiasm and the hypophysis are considered parts of the telencephalon on embryological grounds, but are usually assigned to the diencephalon as a matter of convenience in descriptive anatomy.)

FIG. 56.
THE TELENCEPHALON OR CEREBRUM

1 corpus callosum
2 cerebral hemisphere
3 longitudinal cerebral
 fissure

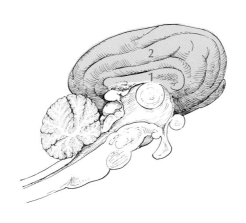
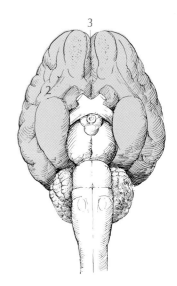

TELENCEPHALON OR CEREBRUM

The cerebrum is the largest portion of the brain. Its surface
is convoluted, and it is divided into bilaterally symmetrical
cerebral hemispheres by the longitudinal cerebral fissure, which
lies in the median sagittal plane. It is composed of a central area
of white matter consisting largely of medullated nerve fibers,
and a peripheral area of gray matter (the cerebral cortex), consisting
largely of cell bodies and nonmedullated processes. The two
cerebral hemispheres are connected by a transverse band of fibers
termed the corpus callosum, and each hemisphere contains one of the
lateral ventricles.

FIG. 57.
THE RHINENCEPHALON AND NEOPALLIUM

1 lateral olfactory stria
2 medial olfactory stria
3 olfactory bulb
4 olfactory tract
5 olfactory trigone
6 piriform lobe

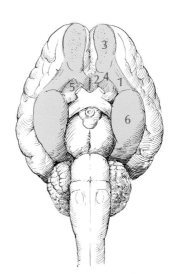
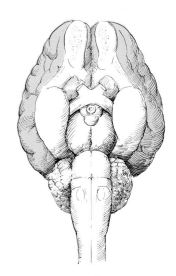

rhinencephalon   neopallium

cerebrum { rhinencephalon { olfactory bulbs
 corpus striatum   olfactory tracts
 neopallium   olfactory striae
          olfactory trigone
          piriform lobe }

The fundamental parts of the cerebrum are the rhinencephalon,
the corpus striatum, and the neopallium.

The rhinencephalon is the olfactory portion of the cerebrum.
It includes the olfactory bulbs, the olfactory tracts, the olfactory striae,
the olfactory trigone, and the piriform lobe.

The neopallium is the nonolfactory portion of the cerebral cortex.

The corpus striatum is a subcortical mass consisting mostly of
gray matter. It is seen as a dark area within the cerebral white
matter when the brain is sectioned (see Fig. 61, p. 80).

FIG. 58.
THE DIENCEPHALON

1 epithalamus
2 habenular trigone
3 hypophysis
4 hypothalamic nucleus
5 hypothalamus
6 infundibulum
7 mammillary body
8 massa intermedia
9 optic chiasm
10 pineal body
11 posterior commissure
12 thalamus

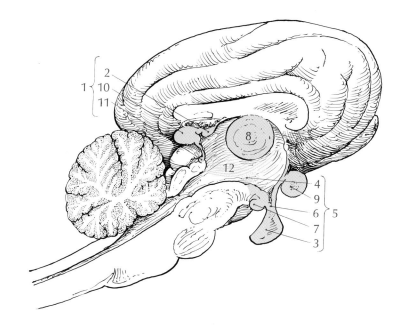

DIENCEPHALON

diencephalon {
  thalamus
  epithalamus { pineal body / posterior commissure / habenular trigone
  hypothalamus { mammillary bodies / tuber cinereum / infundibulum / hypophysis / optic chiasm / hypothalamic nuclei
}

The diencephalon includes the thalamus, the epithalamus, and the hypothalamus.

The thalamus consists of two bilaterally symmetrical ovoid masses of gray matter which lie on either side of the third ventricle and form most of its lateral walls. They are connected by a transverse band of gray matter termed the massa intermedia, which bridges the third ventricle.

The epithalamus includes the pineal body, the posterior commissure, and the habenular trigone.

The hypothalamus forms most of the floor and the lower part of the lateral walls of the third ventricle. It includes the mammillary bodies, the tuber cinereum, infundibulum, hypophysis, and optic chiasm. It also includes a group of cells termed the hypothalamic nuclei, which lie on either side of the lower part of the third ventricle, just under the thalamus.

FIG. 59.
THE MESENCEPHALON

1 aqueduct
2 cerebral peduncle
3 corpora quadrigemina

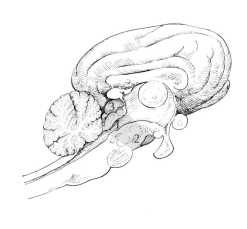

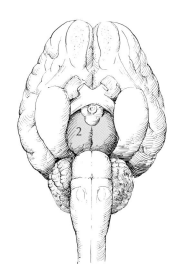

MESENCEPHALON

The mesencephalon is a short segment which connects the rhombencephalon with the prosencephalon. It surrounds the cerebral aqueduct and includes a ventral portion—the cerebral peduncles—and a dorsal portion—the corpora quadrigemina. The cerebral peduncles are bundles of white fibers which pass between the pons and the cerebrum. The corpora quadrigemina are four rounded bulges on the dorsal surface of the mesencephalon. The two anterior bulges are termed the superior colliculi; the two posterior bulges are termed the inferior colliculi.

mesencephalon
- corpora quadrigemina
  - superior colliculi
  - inferior colliculi
- cerebral peduncles

FIG. 60.

THE RHOMBENCEPHALON

1 cerebellum
2 myelencephalon or medulla oblongata
3 pons

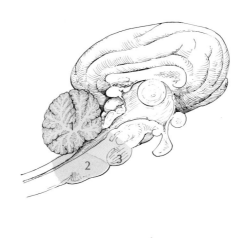
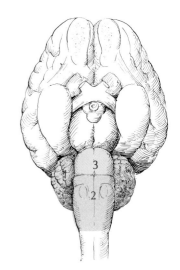

RHOMBENCEPHALON

The rhombencephalon includes the metencephalon and the myelencephalon.

The metencephalon is composed of the pons and the cerebellum.

The pons is seen as a bulge on the ventral surface of the brain stem between the cerebral peduncles and the medulla. It consists of a transverse band of white fibers which arches across the anterior end of the medulla. On each side the fibers form a bundle, the middle cerebellar peduncle, which enters the cerebellum.

The cerebellum lies posterior to the cerebrum and above the fourth ventricle. It consists of a central mass of white matter covered by a layer of gray matter, and its surface is furrowed by numerous curved grooves. In sagittal section the white and gray matter of the cerebellum are seen to form a compact tree-like pattern termed the arbor vitae. The median portion of the cerebellum is termed the vermis; the two lateral portions are termed the cerebellar hemispheres. On each side, three cerebellar peduncles join the white matter of the cerebellum. The posterior cerebellar peduncle connects the cerebellum with the medulla and spinal cord. The anterior cerebellar peduncle passes to the mesencephalon. The middle cerebellar peduncle passes to the pons (see Figs. 67 and 68, pp. 90 and 91).

The myelencephalon, or medulla oblongata, is the portion of the brain stem between the pons and the spinal cord.

rhombencephalon
- metencephalon
  - pons
  - cerebellum
- myelencephalon

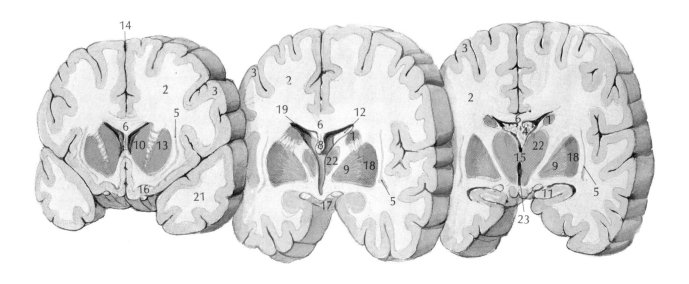

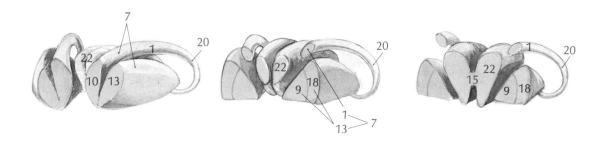

FIG. 61.

THE CORPUS STRIATUM AND THALAMUS

Above, the corpus striatum (blue) and thalamus (red) are seen as exposed in serial cross sections of the human brain. Below, the same structures are reconstructed schematically (adapted from Sobotta and Villiger).

1	caudate nucleus	13	lentiform nucleus
2	central white matter	14	longitudinal cerebral
3	cerebral cortex		fissure
4	choroid plexus	15	massa intermedia
5	claustrum	16	olfactory tract
6	corpus callosum	17	optic chiasm
7	corpus striatum	18	putamen
8	fornix	19	right lateral ventricle
9	globus pallidus	20	tail of caudate nucleus
10	head of caudate nucleus	21	temporal lobe
11	hippocampus	22	thalamus
12	left lateral ventricle	23	third ventricle

The brain and spinal cord are surrounded by three membranes collectively termed the meninges. The strongest and most superficial membrane is the tough, semiopaque dura mater. Because it is removed in commercially prepared sheep brains, it may best be seen in the sagittal section of the cat's head.

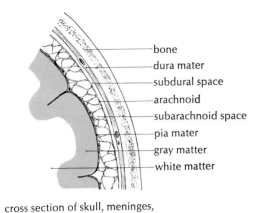

- bone
- dura mater
- subdural space
- arachnoid
- subarachnoid space
- pia mater
- gray matter
- white matter

cross section of skull, meninges, and cerebral cortex

The innermost membrane is the pia mater. It is seen in the sheep brain as the thin, vascular covering which adheres closely to the folds and grooves of the surface of the brain. Between the pia and the dura is the arachnoid, a delicate, transparent membrane which lies close to the dura and does not adhere to the brain surface as the pia mater does. It is difficult to distinguish grossly in the brain of the cat or sheep. The arachnoid is attached to the pia mater by delicate strands of connective tissue. The space between the dura and the arachnoid is the subdural space. The space between the arachnoid and the pia mater is the subarachnoid space, which contains the lymph-like cerebrospinal fluid.

DORSAL ASPECT OF BRAIN

Examine the dorsal aspect of the intact sheep brain. The convolutions of the cerebral cortex are separated from each other by fissures. The fissures are termed sulci, and the elevations between the sulci are termed gyri. Probe the depth of the sulci and observe the way in which the pia mater and the vessels penetrate them. Gently draw the two cerebral hemispheres apart and observe the corpus callosum, which appears as a transverse band of fibers within the longitudinal cerebral fissure, about half an inch below the surface. See the corpus callosum as illustrated in Figure 63, page 85.

The fissure which separates the cerebral hemispheres from the cerebellum is termed the transverse fissure. Pull the cerebrum away from the cerebellum and identify the pineal body and the corpora quadrigemina, which lie on the dorsal surface of the brain stem and are concealed by the cerebrum. Refer to Figure 67, page 90.

Lift the vermis away from the medulla and look between them to observe the cavity of the fourth ventricle, which lies below the vermis. The posterior portion of the roof of the fourth ventricle is termed the medullary velum. It consists of a thin layer of white substance covered by pia mater, and is easily torn when the cerebellum is pulled away from the medulla.

VENTRAL ASPECT OF BRAIN

Referring to Figure 62, examine the ventral surface of the sheep brain. At the anterior end of the brain are the paired olfactory bulbs. In life they lie within the olfactory fossae of the skull. Observe the position of the olfactory bulbs as seen in the sagittal section of the cat's head, and identify the olfactory fossa and the cribiform plate of the ethmoid in the sagittal section of the skull.

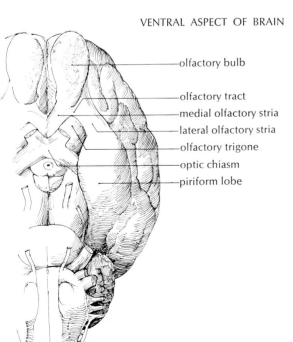

- olfactory bulb
- olfactory tract
- medial olfactory stria
- lateral olfactory stria
- olfactory trigone
- optic chiasm
- piriform lobe

The first cranial nerve (olfactory) consists of numerous fibers which pass from the mucous membrane of the nasal cavity through the holes in the cribiform plate of the ethmoid to the olfactory bulb. Cut the olfactory bulb and observe that it is hollow. The ventricle of the olfactory bulb is connected with the lateral ventricle of the brain via a small canal in the olfactory tract.

Just posterior to the olfactory bulb is the short olfactory tract, which soon divides into medial and lateral olfactory striae. The lateral stria is a clearly defined band which passes laterally to join the piriform lobe; the medial stria, which passes toward the midline and appears to blend with the cerebral cortex, is less clearly defined. The triangular area bounded by the medial and lateral olfactory striae is termed the olfactory trigone. The piriform lobe is the portion of the cerebrum that lies just lateral to the optic chiasm and the cerebral peduncles.

OPTIC CHIASM   At the posterior boundary of the olfactory trigone is the optic chiasm, formed by the crossing of the optic nerves and tracts. Within the area bounded by the optic chiasm and the cerebral peduncles are the mammillary bodies and the tuber cinereum. The tuber cinereum is the slight bulge which lies in the midline immediately posterior to the optic chiasm. The hypophysis, or pituitary gland, is suspended from the tuber cinereum by a tubular stalk termed the infundibulum. In life the hypophysis lies within the sella turcica of the basisphenoid. Identify the hypophysis and the sella turcica in the sagittal section of the cat's head.

MAMMILLARY BODIES   The mammillary bodies are small paired elevations posterior to the tuber cinereum. Each mammillary body consists of a nucleus of gray matter enclosed in a capsule of white matter.

CEREBRAL PEDUNCLES   The cerebral peduncles are seen on the base of the brain as two broad white bands which are connected posteriorly to the pons and which diverge, passing on either side of the mammillary bodies and the tuber cinereum to enter the cerebrum. They are crossed anteriorly by the optic tracts.

Posterior to the mammillary bodies, on the ventral surface of the cerebral peduncle, is the superficial origin of the third cranial nerve (oculomotor). The fourth cranial nerve (trochlear) originates on the dorsal surface of the brain stem, just posterior to the corpora quadrigemina, and curves around the lateral aspect of the cerebral peduncle between the cerebrum and the cerebellum.

Posterior to the cerebral peduncles lies the pons, a transverse band of fibers which arches across the brain stem between the medulla oblongata and the cerebral peduncles. Trace the pons laterally, observing that it extends into the cerebellum as the middle cerebellar peduncle. The fifth cranial nerve (trigeminal) originates from the lateral surface of the pons.

Posterior to the pons is the trapezoid body, a narrow band of transverse fibers which crosses the anterior end of the medulla oblongata. The trapezoid body originates near the acoustic area and passes ventrally and anteriorly, turning forward before it reaches the midline.

In the midventral line of the medulla oblongata is the ventral median fissure, which is flanked by longitudinal tracts termed pyramids.

The sixth cranial nerve (abducens) originates posterior to the pons, near the lateral margin of the pyramid. The seventh cranial nerve (facial) originates between the pons and the trapezoid body, immediately posterior to the origin of the fifth cranial nerve. The eighth cranial nerve (auditory) arises near the point where the trapezoid body passes under the cerebellum. The ninth (glossopharyngeal), tenth (vagus), and eleventh (spinal accessory) cranial nerves arise along the lateral side of the medulla oblongata posterior to the origin of the eighth nerve. The twelfth cranial nerve (hypoglossal) arises by numerous small rootlets near the posterior end of the medulla at the lateral edge of the pyramid.

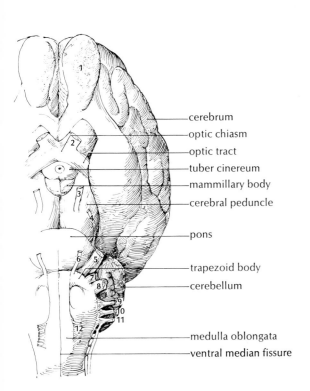

cerebrum
optic chiasm
optic tract
tuber cinereum
mammillary body
cerebral peduncle
pons
trapezoid body
cerebellum
medulla oblongata
ventral median fissure

FIG. 62.
VENTRAL VIEW OF THE SHEEP BRAIN

1   abducens nerve (6)
2   anterior cerebral artery
3   auditory nerve (8)
4   basilar artery
5   cerebellum
6   cerebral peduncle
7   facial nerve (7)
8   glossopharyngeal
    nerve (9)
9   hypoglossal nerve (12)
10 infundibulum
11 internal carotid artery
12 lateral olfactory stria
13 mammillary body
14 medial olfactory stria
15 medulla oblongata
16 middle cerebral artery
17 oculomotor nerve (3)
18 olfactory bulb and
    nerve (1)
19 olfactory tract
20 olfactory trigone
21 optic chiasm
22 optic nerve (2)
23 optic tract
24 pons
25 posterior cerebral artery
26 rhinal fissure
27 spinal accessory nerve (11)
28 spinal cord
29 trapezoid body
30 trigeminal nerve (5)
31 trochlear nerve (4)
32 tuber cinereum
33 vagus nerve (10)

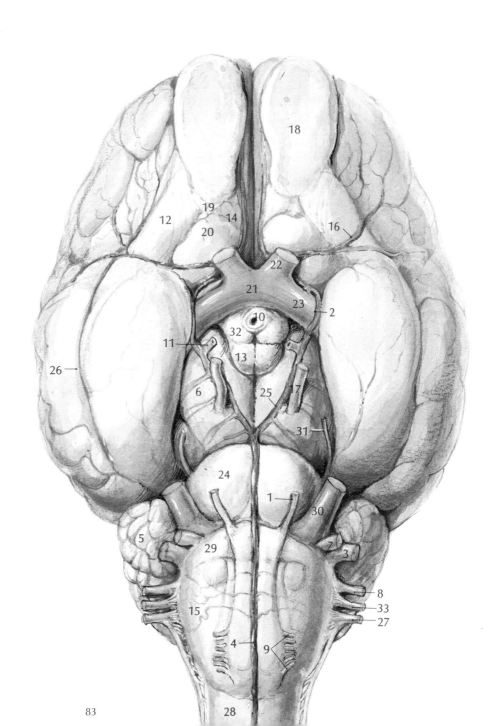

SAGITTAL SECTION OF BRAIN

Referring to Figure 63, cut the sheep brain in sagittal section. Use a long knife and make the cut with a single stroke, keeping it as nearly as possible in the median sagittal plane.

CORPUS CALLOSUM

The corpus callosum as seen in the sagittal section consists of a transverse band of white fibers. The central portion is termed the trunk. The anterior part is the genu; the posterior part is the splenium. The corpus callosum forms the roof of the lateral ventricles.

FORNIX

Ventral to the corpus callosum is the fornix, a bilaterally symmetrical pair of bands composed of white fibers which arch over the thalamus and the third ventricle. Only the medial portion of the fornix is seen in the sagittal section. Between the fornix and the corpus callosum is the septum pellucidum, a thin, vertical membrane consisting of two lamina. The septum pellucidum separates the two lateral ventricles of the brain.

BOUNDARIES OF THIRD VENTRICLE

The sagittal section cuts directly through the third and fourth ventricles, but the cavity of the third ventricle is so narrow that its boundaries may not be readily apparent. The lateral walls of the third ventricle are formed by the thalamus, and the massa intermedia is seen as an oval area of gray matter in the middle of the ventricle.

Anteriorly the third ventricle is bounded by the lamina terminalis, a thin layer of gray matter extending from the optic chiasm to the corpus callosum. The lamina terminalis is crossed by the anterior commissure, a small transverse band of white fibers which joins the two cerebral hemispheres.

The floor of the third ventricle consists of structures belonging to the hypothalamus: the optic chiasm, tuber cinereum, infundibulum, and mammillary bodies, all of which were observed in the ventral view of the brain.

Posteriorly the third ventricle narrows to form a small channel, the cerebral aqueduct, which is bordered by the cerebral peduncles ventrally and by the corpora quadrigemina dorsally. Above the opening of the cerebral aqueduct the third ventricle is bounded by the structures composing the epithalamus: the pineal body, the posterior commissure, and the habenular trigone. The habenular trigone is a small triangular area anterior to the superior colliculus; it is best seen in the dorsal view of the brain stem (see Fig. 67, p. 90). It contains paired nuclei which are connected by the habenular commissure. The posterior commissure is a band of white fibers which connects the two cerebral hemispheres. The pineal body is a rounded structure which lies between the corpora quadrigemina and the splenium of the corpus callosum.

The roof of the third ventricle is formed by a thin layer of epithelium, which is continuous with the epithelial lining of the third and lateral ventricles. It is covered by a layer of pia mater termed the tela choroidea.

LATERAL VENTRICLES

Trim away the septum pellucidum and probe the lateral ventricle. Pass a probe through the foramen of Monro and confirm the fact that it forms a passage between the third and lateral ventricles. Within the lateral ventricle is a delicate membrane termed the choroid plexus. It is continuous with the tela choroidea and consists of a highly vascular, tufted extension of the pia mater. Choroid

CHOROID PLEXUS

plexuses are also present in the third and fourth ventricles.

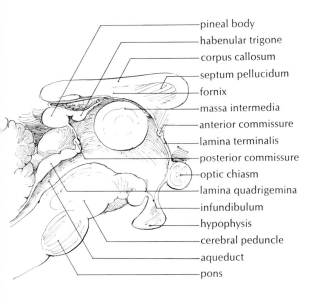

pineal body
habenular trigone
corpus callosum
septum pellucidum
fornix
massa intermedia
anterior commissure
lamina terminalis
posterior commissure
optic chiasm
lamina quadrigemina
infundibulum
hypophysis
cerebral peduncle
aqueduct
pons

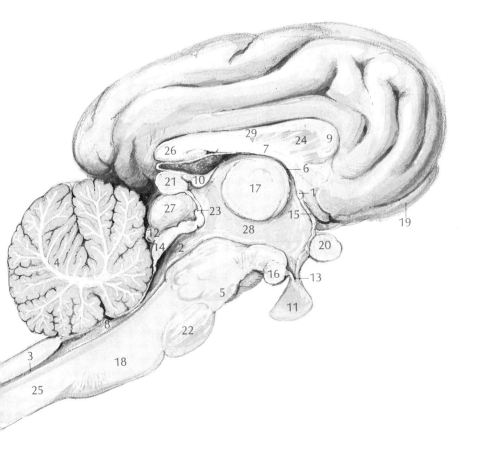

1	anterior commissure
2	aqueduct
3	central canal
4	cerebellum
5	cerebral peduncle
6	foramen of Monro
7	fornix
8	fourth ventricle
9	genu of corpus callosum
10	habenular trigone
11	hypophysis
12	inferior colliculus
13	infundibulum
14	lamina quadrigemina
15	lamina terminalis
16	mammillary body
17	massa intermedia
18	medulla oblongata
19	olfactory bulb
20	optic chiasm
21	pineal body
22	pons
23	posterior commissure
24	septum pellucidum
25	spinal cord
26	splenium
27	superior colliculus
28	third ventricle
29	trunk of corpus callosum

CRANIAL NERVES   Twelve cranial nerves are attached to the base of the brain and exit from the cranial cavity via various foramina of the skull. The superficial origins of the cranial nerves were seen during the examination of the ventral aspect of the brain. The foramina through which they exit are illustrated in Figures 4-7 on pages 6-9. Because of the small size and the complexity of the structures involved, a detailed dissection of the cranial nerves will not be attempted.

OLFACTORY NERVE   1. The olfactory nerve (sensory) consists of numerous small bundles of nonmedullated fibers which pass from the olfactory bulb through the foramina in the cribiform plate of the ethmoid bone to the mucous membrane of the nasal cavity.

OPTIC NERVE   2. The optic nerve (sensory) arises from the optic chiasm and passes through the optic foramen to reach the posteromedial part of the eyeball. It pierces the sclera and the choroid to supply the retina.

OCULOMOTOR NERVE   3. The oculomotor nerve (motor) arises from the ventral surface of the cerebral peduncle. It passes through the orbital fissure to supply the following eye muscles: levator palpebrae, superior rectus, medial rectus, inferior rectus, and inferior oblique.

TROCHLEAR NERVE   4. The trochlear nerve (motor) arises just posterior to the corpora quadrigemina, curves around the lateral aspect of the cerebral peduncle, and enters the orbit through the orbital fissure. It supplies the superior oblique muscle.

TRIGEMINAL NERVE   5. The trigeminal is the largest of the cranial nerves. It is the sensory nerve to the skin of the face, the mucous membranes, and other internal structures of the head, and it is the motor nerve to the muscles of mastication. It originates from the lateral surface of the pons by a large sensory root and a smaller, more ventrally placed, motor root. Near its origin the sensory root enlarges to form the semilunar ganglion, which gives off three large branches: the ophthalmic nerve, the maxillary nerve, and the mandibular nerve.

OPHTHALMIC NERVE   The ophthalmic nerve (sensory) passes through the orbital fissure. Within the orbit it divides into several branches which supply the eyeball, forehead, and nose.

MAXILLARY NERVE   The maxillary nerve (sensory) passes through the foramen rotundum and divides into several branches which are distributed to the palate, upper teeth, and upper lip, and part of the forehead and cheek.

MANDIBULAR NERVE   The motor root of the trigeminal bypasses the semilunar ganglion and joins the third branch of the ganglion to form the mandibular nerve, which is therefore both sensory and motor. The mandibular nerve exits through the foramen ovale and divides into several branches. Motor components are distributed chiefly to the muscles of mastication; sensory components go to the temporal region, ear, cheek, jaw, lower teeth, and tongue.

ABDUCENS NERVE   6. The abducens nerve (motor) originates posterior to the pons, near the lateral margin of the pyramid. It passes through the orbital fissure and supplies the external rectus and retractor oculi.

FACIAL NERVE   7. The facial nerve (motor and sensory) originates between the pons and the trapezoid body, just posterior to the origin of the trigeminal nerve. It enters the internal auditory meatus, passes through the facial canal of the petrous bone, and emerges at the stylomastoid foramen. It supplies motor components to most of the muscles of the head, except those of mastication, and sensory components to the tongue and soft palate.

AUDITORY NERVE   8. The auditory nerve conveys sensation from the inner ear. It originates from the medulla near the point where the trapezoid body passes under the cerebellum. It enters the internal auditory meatus and divides into cochlear and vestibular nerves. The cochlear nerve transmits auditory sensations from the cochlea. The vestibular nerve transmits sensations of equilibrium from the semicircular canals, the sacculus, and the utriculus.

GLOSSOPHARYNGEAL NERVE   9. The glossopharyngeal nerve (sensory and motor) originates at the lateral side of the medulla, just posterior to the auditory nerve. It passes through the jugular foramen, over the tympanic bulla, and divides into two branches. One branch goes to the muscles and mucosa of the pharynx; the other branch goes to the tongue. The glossopharyngeal nerve is motor to the pharyngeal muscles, and sensory to the tongue and pharynx.

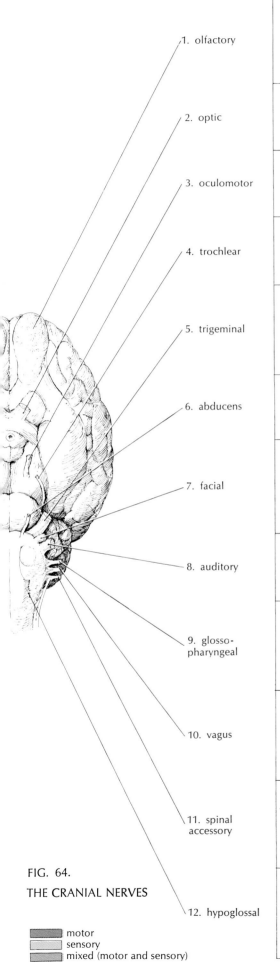

1. olfactory

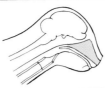

Sensory. From olfactory bulb via cribiform plate to olfactory mucosa.

2. optic

Sensory. From optic chiasm via optic foramen to retina.

3. oculomotor

Motor. From cerebral peduncle via orbital fissure to levator palpebrae, superior rectus, medial rectus, inferior rectus, and inferior oblique.

4. trochlear

Motor. From dorsal surface of brain stem via orbital fissure to superior oblique.

5. trigeminal

Motor. From medulla via orbital fissure to lateral rectus and retractor oculi.

Mixed. From lateral surface of pons. 3 branches:

1. Ophthalmic. Sensory. From semilunar ganglion via orbital fissure to eye, forehead, and nose.

2. Maxillary. Sensory. From semilunar ganglion via foramen rotundum to palate, upper teeth and lip, forehead, and cheek.

3. Mandibular. From semilunar ganglion via foramen ovale. Sensory to ear, cheek, jaw, tongue, and mouth; motor to muscles of mastication.

6. abducens

7. facial

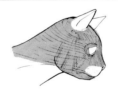

Mixed. From caudal margin of pons via internal auditory meatus and stylomastoid foramen. Motor to muscles of head; sensory to lingual mucosa.

8. auditory

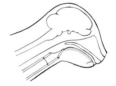

Sensory. From medulla via internal auditory meatus to membranous labyrinth.

9. glosso-pharyngeal

Mixed. From medulla via jugular foramen. Sensory to tongue and pharyngeal mucosa; motor to pharyngeal muscles.

10. vagus

Mixed. From medulla via jugular foramen. Sensory to external ear and laryngeal mucosa. Motor to muscles of pharynx and larynx. Mixed to heart, lungs, and abdominal viscera.

11. spinal accessory

Motor. From medulla and spinal cord via jugular foramen to cleidomastoid, sternomastoid, and trapezius muscles.

12. hypoglossal

Motor. From medulla via hypoglossal foramen to hyoid and tongue muscles.

FIG. 64.

THE CRANIAL NERVES

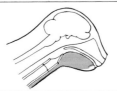

motor
sensory
mixed (motor and sensory)

10. The vagus nerve (sensory and motor) originates at the lateral side of the medulla just posterior to the glossopharyngeal nerve, and passes through the jugular foramen. Near its origin are two ganglia: the jugular ganglion, which lies within the jugular foramen, and the nodose ganglion, which lies just beyond the jugular foramen, near the superior cervical sympathetic ganglion. The vagus passes posteriorly, lying lateral to the common carotid artery and bound closely to the cervical portion of the sympathetic trunk. The cervical portion of the vagus gives small branches to the external ear, the pharynx, and the larynx.

The vagus nerve and the sympathetic trunk separate just before entering the thorax. Within the thorax the right vagus lies along the trachea, and the left vagus lies along the esophagus. Near the first rib the vagus receives communicating branches from the sympathetic trunk. Near the aortic arch the left vagus gives rise to the left recurrent laryngeal nerve, which curves around the aortic arch and passes anteriorly, lying along the lateral side of the trachea, to reach the larynx. The right recurrent laryngeal nerve arises from the right vagus. It curves around the right subclavian artery and passes forward along the trachea to reach the larynx.

Near the root of the lung the vagus gives rise to the cardiac and pulmonary plexuses. Many branches pass from these plexuses to the heart, lungs, and associated structures. Posteriorly the vagus nerve of each side is represented by dorsal and ventral branches which lie along the esophagus, to which they give branches. Near the roots of the lungs the ventral branches of the right and left vagus nerves unite to form the ventral division of the vagus. The dorsal branches of the right and left vagus nerves unite near the diaphragm to form the dorsal division of the vagus.

The ventral division passes through the esophageal opening of the diaphragm, lying on the ventral aspect of the esophagus, and ramifies on the lesser curvature of the stomach. The dorsal division passes through the esophageal opening of the diaphragm, lying on the dorsal aspect of the esophagus, and ramifies on the greater curvature of the stomach. On the stomach the dorsal and ventral divisions of the vagus form dorsal and ventral gastric plexuses, which anastomose with each other and with the celiac plexus. Vagus fibers extend through the celiac plexus to supply the abdominal viscera as far as the transverse colon.

11. The spinal accessory nerve (motor) is formed by numerous rootlets which arise from the lateral surface of the spinal cord and medulla. It passes through the jugular foramen, where it is united with the vagus, sympathetic, and hypoglossal nerves by numerous fine branches which form the pharyngeal plexus. It pierces the cleidomastoid muscle, which it supplies, and divides into two branches. One supplies the sternomastoid; the other supplies the trapezius muscles.

12. The hypoglossal nerve (motor) originates by numerous rootlets from the ventral side of the medulla. It passes through the hypoglossal foramen, lying lateral to the vagus nerve, the sympathetic trunk, and the common carotid artery. It gives branches to certain ventral neck muscles and to the muscles of the tongue.

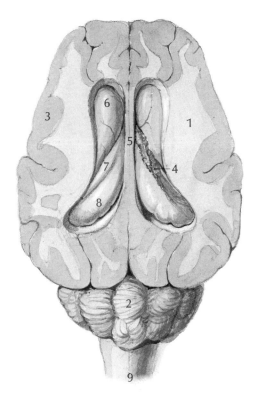

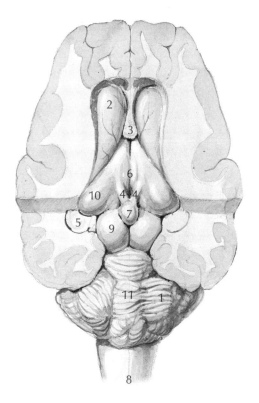

FIG. 65.

DORSAL VIEW OF THE SHEEP
BRAIN, SUPERFICIAL DISSECTION

1 central white matter
2 cerebellum
3 cerebral cortex
4 choroid plexus
5 corpus callosum
6 corpus striatum
7 fornix
8 hippocampus
9 spinal cord

FIG. 66.

DORSAL VIEW OF THE SHEEP
BRAIN, DEEP DISSECTION

1 cerebellar hemisphere
2 corpus striatum
3 fornix (cut)
4 habenular trigone
5 hippocampus
6 massa intermedia
7 pineal body
8 spinal cord
9 superior colliculus
10 thalamus
11 vermis

DISSECTION OF CEREBRUM   For the dissection of the cerebrum and brain stem, as described
below, a new sheep brain should be used if possible. If new
brains are not supplied, use the brain you have cut in sagittal
section, dissecting only one side of it and keeping the other for
reference.

Starting on the dorsal aspect of the cerebrum, make a series of
thin sections, shaving away successive layers until you reach the
lateral ventricles. Then use scissors to trim away the tissue above
the lateral ventricles, making a dissection similar to Figure 65.
The caudate nucleus, fornix, and hippocampus are seen in the
floor of the body and anterior horn of the lateral ventricle. The
caudate nucleus is part of the corpus striatum. The fornix, a band
of white fibers which arches over the thalamus, was seen in
the sagittal section of the brain. The hippocampus is an elevation in
the floor of the lateral ventricle. The choroid plexus is seen
as a strand of vascular tissue lying dorsal to the fornix. Lift the

FIG. 67.

DORSAL VIEW OF THE BRAIN STEM

1   acoustic area
2   acoustic striae
3   anterior cerebellar
    peduncle
4   clava
5   corpora quadrigemina
6   cut surface of thalamus
    and corpus striatum
7   dorsal median sulcus
8   dorsolateral sulcus
9   fasciculus cuneatus
10 fasciculus gracilis
11 fourth ventricle
12 habenular trigone
13 inferior colliculus
14 massa intermedia
15 medial geniculate body
16 middle cerebellar
    peduncle
17 opening of aqueduct
18 opening of central canal
19 pineal body
20 posterior cerebellar
    peduncle
21 superior colliculus
22 thalamus
23 third ventricle
24 trochlear nerve
25 tuberculum cuneatum

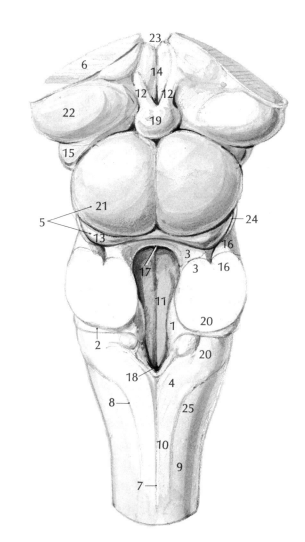

hippocampus away from the brain stem, observing that the thalamus lies ventral and medial to it.

Trim away the corpus callosum, cut the fornix, and remove the posterior portion of the cerebral hemisphere to expose the thalamus, making a dissection similar to Figure 66. Identify the structures illustrated. Also study the fornix, hippocampus, corpus striatum, and thalamus in cross sections as seen in a demonstration dissection.

DISSECTION OF BRAIN STEM   Cut through the brain stem anterior to the optic chiasm and remove the cerebrum. Remove the cerebellum by cutting the cerebral peduncles. The remaining portion of the brain is the brain stem. Examine the dorsal aspect of the brain stem and identify the structures illustrated in Figure 67.

Within the fourth ventricle is an elevation termed the acoustic area. From it a band of fibers (acoustic striae) passes over the posterior cerebellar peduncle to the superficial origin of the auditory nerve.

FIG. 68.

LATERAL VIEW OF THE BRAIN STEM

1 abducens nerve (6)
2 acoustic striae
3 anterior cerebellar
   peduncle
4 auditory nerve (8)
5 cerebral peduncle
6 clava
7 cut surface of thalamus
   and corpus striatum
8 facial nerve (7)
9 glossopharyngeal nerve (9)
10 hypoglossal nerve (12)
11 inferior colliculus
12 infundibulum
13 lateral geniculate body
14 mammillary body
15 medial geniculate body
16 middle cerebellar
   peduncle
17 oculomotor nerve (3)
18 optic chiasm
19 optic tract
20 pineal body
21 pons
22 posterior cerebellar
   peduncle
23 spinal accessory nerve (11)
24 superior colliculus
25 thalamus
26 trapezoid body
27 trigeminal nerve (5)
28 trochlear nerve (4)
29 tuberculum cuneatum
30 vagus nerve (10)

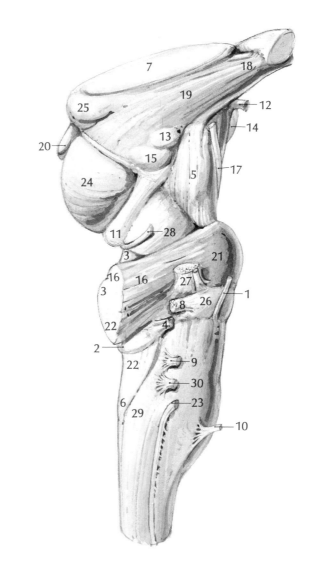

Identify the structures illustrated in Figure 68. Compare the lateral view of the brain stem with the dorsal view, the ventral view, and the sagittal section.

SPINAL CORD The spinal cord is that portion of the nervous system which lies within the vertebral canal. Anteriorly it is continuous with the medulla oblongata. It extends from the foramen magnum to the caudal region, and has the form of a cylinder which is somewhat flattened dorsoventrally.

Those portions of the cord which give rise to the nerves innervating the forelimbs and the hindlimbs are somewhat wider than the rest of the cord and are termed the cervical and lumbar enlargements. The cervical enlargement lies between the fourth cervical and the first thoracic vertebrae; the lumbar enlargement lies between the third and seventh lumbar vertebrae. Between the cervical and lumbar enlargements the spinal cord is of fairly uniform diameter, but posterior to the lumbar enlargement it tapers to a conical tip termed the conus medullaris. The conus medullaris is continuous with a slender strand termed the filum terminale, which extends into the caudal region.

A minute passage termed the central canal runs throughout the length of the spinal cord. It is continuous anteriorly with the fourth ventricle. Posteriorly it extends to the filum terminale.

A groove, the ventral median fissure, runs along the ventral surface of the spinal cord in the midline. On the dorsal surface of the cord is a similar but somewhat more shallow groove, the dorsal median sulcus. Parallel with and lateral to the dorsal median sulcus is another groove, the dorsolateral sulcus, along which the dorsal roots of the spinal nerves are attached to the cord.

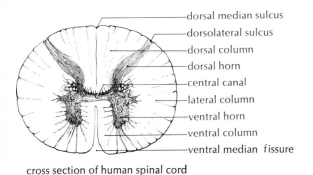

dorsal median sulcus
dorsolateral sulcus
dorsal column
dorsal horn
central canal
lateral column
ventral horn
ventral column
ventral median fissure

cross section of human spinal cord
(after Gray)

SPINAL NERVES

The membranes that invest the spinal cord are continuous with those that invest the brain. The cerebral dura mater is a single membrane representing a fusion of the dura of lower forms and the periosteum of the skull. These two layers separate at the foramen magnum; the outer layer constitutes the periosteum of the vertebral canal, and the inner layer forms a tubular sheath which loosely invests the spinal cord throughout its length. The pia mater and arachnoid of the spinal cord are similar to the pia mater and arachnoid of the brain. The central canal and the subdural space of the spinal cord contain cerebrospinal fluid.

In cross section the cord is seen to consist of a peripheral area of white matter which surrounds a central, H-shaped area of gray matter. The white matter is divided into three pairs of columns, or funiculi, as labeled in the marginal diagram.

Thirty-eight pairs of spinal nerves arise from the spinal cord. They are named according to the regions of the vertebral canal through which they exit: eight cervical, thirteen thoracic, seven lumbar, three sacral, and seven or eight caudal nerves. The first cervical nerve exits from the vertebral canal through the atlantal foramen. The second cervical nerve exits between the arches of the axis and atlas. All the other spinal nerves exit via intervertebral foramina. Each spinal nerve is enclosed near its exit by a tubular sheath formed by an extension of the meninges.

ROOTS OF SPINAL NERVES

Each nerve arises by a dorsal (sensory) and a ventral (motor) root. The dorsal root originates as a number of separate nerve bundles which arise from the dorsolateral groove. Just proximal to its union with the ventral root it bears a spinal ganglion, an oval swelling which contains the cell bodies of the sensory fibers. The ventral root consists of motor fibers having cell bodies within the spinal cord. It arises by a number of small rootlets, which converge and join the dorsal root just distal to the spinal ganglion.

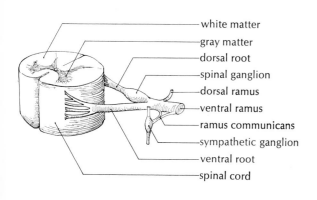

white matter
gray matter
dorsal root
spinal ganglion
dorsal ramus
ventral ramus
ramus communicans
sympathetic ganglion
ventral root
spinal cord

In the cervical region the spinal nerves originate near the intervertebral foramina through which they exit. Posteriorly the distance between the origin and the exit of the spinal nerves becomes greater, so that the roots of the spinal nerves below the cervical region pass posteriorly to reach the intervertebral foramina and assume an increasingly oblique direction. In the sacral and caudal region the roots of the nerves within the vertebral canal are almost vertical, and are collectively termed the cauda equina.

BRANCHES OF SPINAL NERVES

Just outside the intervertebral foramina, the spinal nerve divides into ventral and dorsal rami. The dorsal rami are for the most part much smaller than the ventral rami; they supply the skin and muscles of the dorsal part of the neck and trunk. The ventral rami are distributed to the limbs and to the muscles and skin of the ventral part of the trunk. Each spinal nerve is connected to a ganglion of the sympathetic chain by one or more slender branches termed rami communicantes.

EXPOSURE OF SPINAL CORD   Remove the muscles dorsal to the vertebral column and use small bone clippers to cut through the laminae and remove the neural arches of several vertebrae in the lower thoracic or upper lumbar region. Expose the spinal cord and roots of one or two spinal nerves. Also see a demonstration dissection in which the entire spinal cord is exposed.

CERVICAL NERVES   There are eight pairs of cervical nerves. The first cervical nerve exits from the vertebral canal via the atlantal foramen, the second exits between the arches of the atlas and the axis, and the others exit via intervertebral foramina. The dorsal rami of the cervical nerves supply the muscles and skin on the back of the neck. They are small and will not be seen from the ventral view. The ventral rami are considerably larger than the dorsal rami. They lie on either side of the neck, lateral to the longus capitis muscle, and are easily seen when the sternomastoid and clavotrapezius are cut and pulled back.

CERVICAL PLEXUS   The ventral rami of the first four cervical nerves are loosely connected by intercommunicating branches to form the cervical plexus. These nerves supply the lateral and ventral muscles of the neck.

BRACHIAL PLEXUS   The ventral rami of cervical nerves 5-8 and the first thoracic nerve are united with each other by a complex network of communicating branches and constitute the brachial plexus. Some variation will be found in the pattern of the brachial plexus in different specimens.

EXPOSURE OF BRACHIAL PLEXUS   Review and identify the muscles of the shoulder and upper forelimb. Referring to Figure 69, clear away any remaining vessels, connective tissue, and fat to expose and identify the nerves illustrated.

CERVICAL NERVES   The first cervical nerve is quite small. It passes ventrally on the longus capitis muscle to supply the ventral neck muscles. It gives small communicating branches to the hypoglossal, the vagus, and the sympathetic trunk.

The ventral ramus of the second cervical nerve is large and easily identified near the insertion of the levator scapulae ventralis. It gives branches to the first and third cervical nerves and to the spinal accessory nerve. It then passes between the sternomastoid and the clavotrapezius and divides into several branches, which are distributed to the lateral aspect of the neck and face.

The third and fourth cervical nerves supply the skin and muscles of the neck and shoulder. The fifth cervical nerve is distributed to the muscles and skin of the neck.

PHRENIC NERVE   The fifth and sixth cervical nerves give off slender branches which unite to form the phrenic nerve. In the neck the phrenic lies next to the vagus nerve. On entering the thorax it separates from the vagus and passes ventral to the root of the lung to reach the diaphragm. It is the motor nerve of the diaphragm.

SUPRASCAPULAR NERVE   The suprascapular nerve originates from the sixth, and sometimes the seventh, cervical nerves. It accompanies the transverse scapular artery and passes between the supraspinatus and the subscapularis to supply the supraspinatus and the infraspinatus. Near the shoulder joint it gives off a small superficial branch to the skin of the upper forelimb.

VENTRAL THORACIC NERVES   The anterior ventral thoracic nerve originates from the seventh cervical nerve and passes to the pectoral muscles in company with the ventral thoracic artery and vein. The posterior ventral thoracic nerve arises from the eighth cervical and first thoracic nerves and passes to the pectoral muscles in company with the long thoracic artery and vein.

SUBSCAPULAR NERVES   The first subscapular nerve arises from the sixth and seventh cervical nerves and supplies the subscapularis muscle, which it enters in common with the subscapular artery and vein. It is usually represented by two strands lying close together. The second subscapular nerve arises mainly from the seventh cervical nerve and supplies the teres major. The third subscapular nerve arises from the seventh and eighth cervical nerves and supplies the latissimus dorsi, which it enters with the thoracodorsal artery and vein.

AXILLARY NERVE   The axillary nerve originates from the sixth and seventh cervical nerves and passes between the teres major and subscapular muscles with the posterior humeral circumflex artery and vein. It emerges on the lateral side of the shoulder between the long and lateral heads of the triceps, and supplies some of the lateral shoulder muscles.

LONG THORACIC NERVE   The long thoracic nerve arises from the seventh cervical, passes deep to the scalenes, and lies along the lateral surface of the serratus ventralis, which it supplies.

MUSCULOCUTANEOUS NERVE   The musculocutaneous nerve arises from the sixth and seventh cervical nerves. It divides near the biceps; the anterior branch supplies the biceps and the coracobrachialis, and the posterior branch passes along the medial margin of the biceps to supply the brachialis. Near the elbow it passes dorsal to the biceps to supply the skin of the lower forelimb.

RADIAL NERVE   The radial nerve originates from the seventh and eighth cervical and the first thoracic nerves. It pierces the middle head of the triceps in company with the deep brachial artery, curves around the humerus, and divides into deep and superficial branches which extend into the lower forelimb. It supplies the triceps, the supinator, and the extensor muscles of the lower forelimb.

MEDIAN NERVE   The median nerve arises from the seventh and eighth cervical and the first thoracic nerves. In the upper forelimb it lies along the brachial artery, passing with it through the supracondyloid foramen of the humerus. It supplies the pronators and the flexors of the lower forelimb, with the exception of the flexor carpi ulnaris.

ULNAR NERVE   The ulnar nerve arises from the eighth cervical and first thoracic nerves and lies medial to the brachial artery in the upper forelimb. It curves around the medial epicondyle of the humerus and extends into the lower forelimb, supplying the flexor carpi ulnaris and the ulnar head of the flexor digitorum profundus.

MEDIAL CUTANEOUS NERVE   The medial cutaneous nerve originates from the first thoracic nerve. It supplies the skin on the ulnar side of the lower forelimb.

THORACIC NERVES   The first thoracic nerve contributes to the brachial plexus, as described above. The ventral rami of all the other thoracic nerves

94

FIG. 69.

THE BRACHIAL PLEXUS

1   axillary nerve
2   biceps
3   clavotrapezius
4   common carotid artery
5   coracobrachialis
6   cervical nerve
7   esophagus
8   first subscapular nerve
9   first thoracic nerve
10 hypoglossal nerve (12)
11 innominate artery
12 levator scapulae ventralis
13 long thoracic nerve
14 longus capitis
15 medial cutaneous nerve
16 median nerve
17 musculocutaneous nerve
18 phrenic nerve
19 precava
20 radial nerve
21 scalenes
22 second subscapular
    nerve
23 serratus ventralis
24 spinal accessory nerve (12)
25 splenius
26 subscapularis
27 superior cervical and
    nodose ganglia
28 suprascapular nerve
29 teres major
30 third subscapular nerve
31 thyroid gland
32 trachea
33 triceps
34 ulnar nerve
35 vagus nerve (10)
36 ventral thoracic nerves

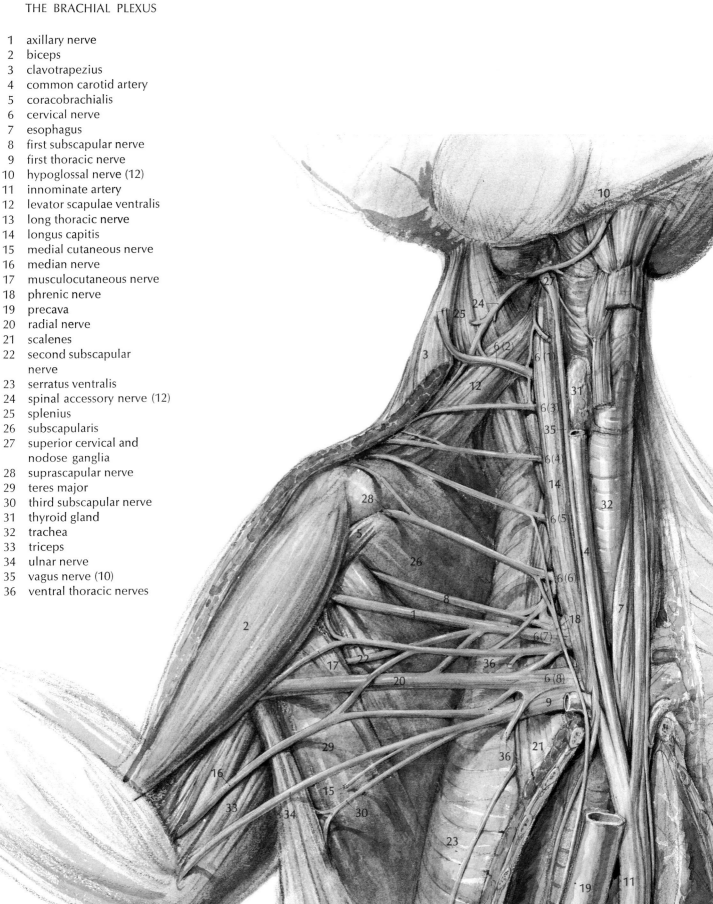

INTERCOSTAL NERVES are termed intercostal nerves. They pass between the intercostal muscles in company with the intercostal artery and vein. Remove the internal intercostal muscle between two adjacent ribs and observe the relations of the intercostal artery, vein, and nerve. The intercostal nerves give branches to the intercostal muscles, to the superficial muscles of the lateral thoracic wall, and to the transversus costarum and the rectus abdominis.

The dorsal rami of the intercostal nerves are small and supply the skin and muscles of the back.

EXPOSURE OF LUMBOSACRAL PLEXUS Remove the abdominal vessels and cut the iliopsoas and the psoas minor. Clear away fat, connective tissue, and muscles to make a dissection similar to Figure 70.

LUMBAR NERVES There are seven pairs of lumbar nerves. The dorsal rami of the lumbar and the sacral nerves are similar to the dorsal rami of the intercostal nerves.

The ventral rami of the first three lumbar nerves divide near their origins into lateral and medial branches which supply the skin and muscles of the lateral and ventral abdominal wall. The ventral rami of the last four lumbar nerves and the three sacral nerves are united by intercommunicating branches to form the lumbosacral plexus.

GENITOFEMORAL NERVE The fourth lumbar nerve is connected with the fifth, and gives rise to the genitofemoral nerve and to the lateral cutaneous nerve. The genitofemoral nerve (illustrated in Fig. 50 on p. 71) is very slender. It lies along the medial surface of the iliopsoas. It crosses the ventral surface of the external iliac artery and vein, follows the deep femoral artery for a short distance, and ramifies on the medial surface of the thigh. In some specimens the genitofemoral gives off a lateral branch that pierces the ventral surface of the psoas minor and passes posteriorly to the anteromedial surface of the thigh and the adjacent abdominal wall.

LATERAL CUTANEOUS NERVE The lateral cutaneous nerve originates from the fourth and fifth lumbar nerves. It passes laterally between the iliopsoas and the psoas minor, accompanying the iliolumbar artery and vein, and extends to the lateral surface of the thigh. It supplies the skin of the thigh.

FEMORAL NERVE The femoral nerve is formed by the fifth and sixth lumbar nerves. It passes posteriorly across the ventral surface of the iliopsoas, to which it gives branches, and pierces the abdominal wall. It gives one or more branches to the sartorius. Another branch passes between the rectus femoris and the vastus medialis in company with the lateral femoral circumflex artery and vein. The femoral nerve continues as the saphenous nerve, which accompanies the femoral artery and vein as far as the ankle, supplying the skin on the medial aspect of the leg.

OBTURATOR NERVE The obturator nerve originates from the sixth and seventh lumbar nerves. It passes posteriorly along the lateral wall of the pelvis, through the obturator foramen, and divides into several branches which supply the adductors, pectineus, and gracilis.

EXPOSURE OF SACRAL NERVES Remove the muscles and fascia of the pelvis as necessary to trace the sacral nerves from the medial aspect. Your dissection should resemble Figure 49 on page 70. Identify the structures illustrated.

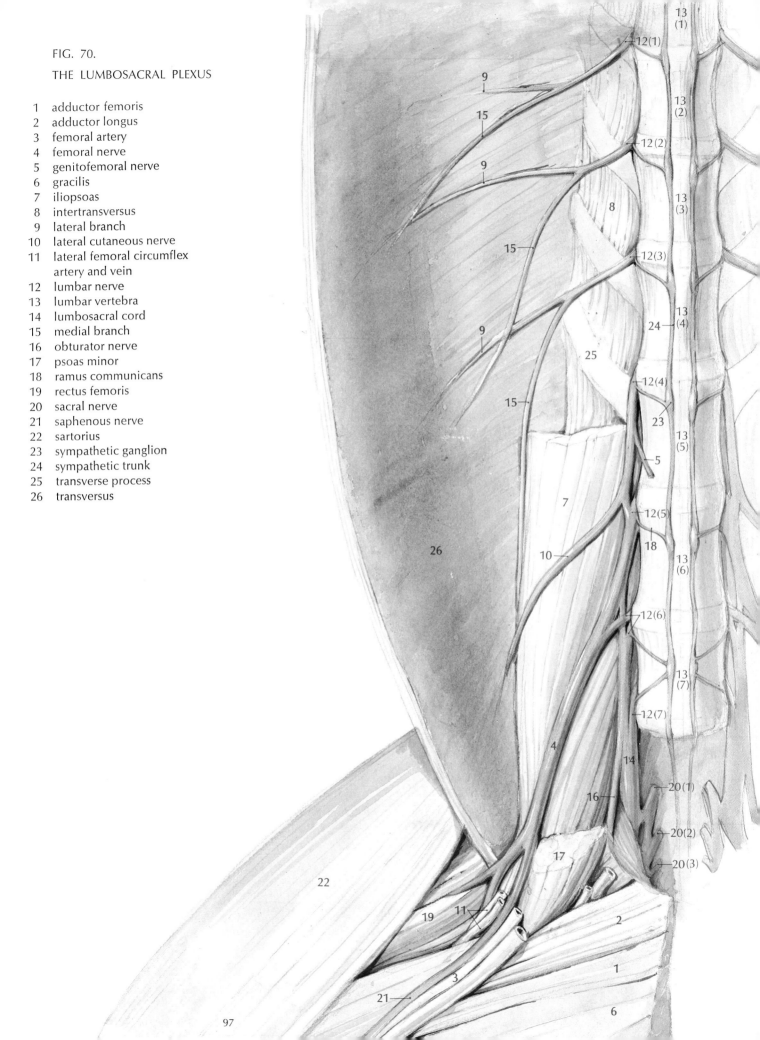

FIG. 70.

THE LUMBOSACRAL PLEXUS

1 adductor femoris
2 adductor longus
3 femoral artery
4 femoral nerve
5 genitofemoral nerve
6 gracilis
7 iliopsoas
8 intertransversus
9 lateral branch
10 lateral cutaneous nerve
11 lateral femoral circumflex
    artery and vein
12 lumbar nerve
13 lumbar vertebra
14 lumbosacral cord
15 medial branch
16 obturator nerve
17 psoas minor
18 ramus communicans
19 rectus femoris
20 sacral nerve
21 saphenous nerve
22 sartorius
23 sympathetic ganglion
24 sympathetic trunk
25 transverse process
26 transversus

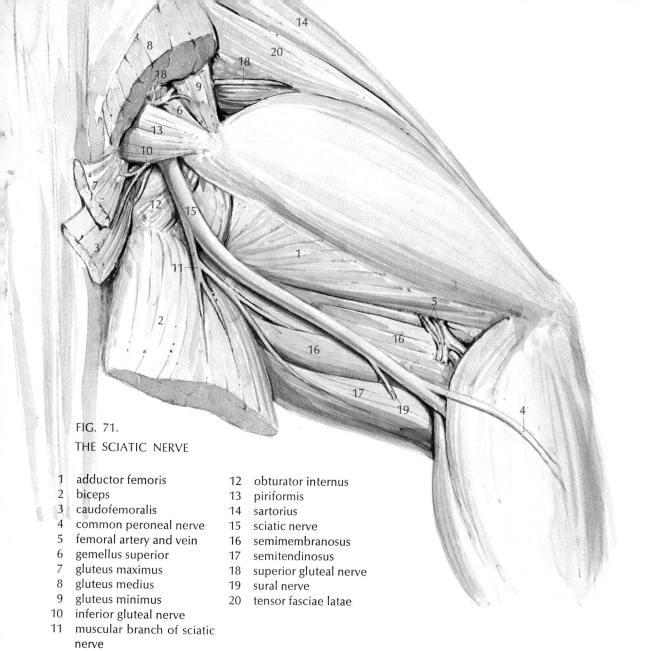

FIG. 71.

THE SCIATIC NERVE

1	adductor femoris	12	obturator internus
2	biceps	13	piriformis
3	caudofemoralis	14	sartorius
4	common peroneal nerve	15	sciatic nerve
5	femoral artery and vein	16	semimembranosus
6	gemellus superior	17	semitendinosus
7	gluteus maximus	18	superior gluteal nerve
8	gluteus medius	19	sural nerve
9	gluteus minimus	20	tensor fasciae latae
10	inferior gluteal nerve		
11	muscular branch of sciatic nerve		

LUMBOSACRAL CORD   A stout band termed the lumbosacral cord passes posteriorly from the sixth and seventh lumbar nerves, connecting them with the SACRAL NERVES three sacral nerves. The lumbosacral cord and the first sacral nerve join to form the sciatic nerve and the superior and inferior gluteal nerves. The sciatic also receives a small branch from the second sacral nerve. A branch connects the first and second sacral nerves. The second and third sacral nerves unite and give rise to the pudendal, posterior femoral cutaneous, and inferior hemorrhoidal nerves.

DISSECTION OF THIGH   Cut through the middle of the biceps femoris. Leave the proximal half of the muscle intact and remove the distal half. Observe and remove the tenuissimus, a slender muscle which lies parallel to and posterior to the sciatic nerve. Remove the tensor fasciae latae. Cut the caudofemoralis, the gluteus maximus, and the gluteus medius to expose the underlying structures. Your dissection should now resemble Figure 71.

SCIATIC NERVE   The sciatic nerve passes out of the pelvis between the ilium and the caudal vertebrae. Near this point it gives a large muscular

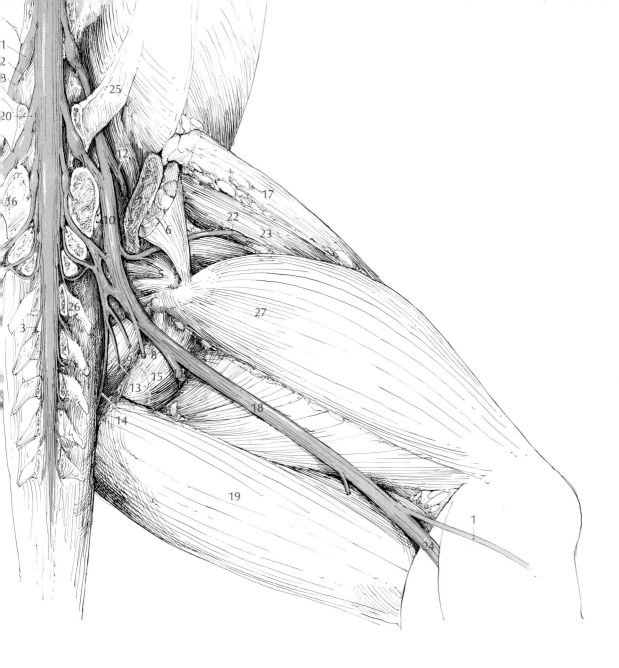

FIG. 72.

THE SACRAL PLEXUS

1	common peroneal nerve	15	quadratus femoris
2	dorsal ramus of sixth lumbar nerve	16	sacrum
3	first coccygeal nerve	17	sartorius
4	first sacral nerve	18	sciatic nerve
5	gemellus superior	19	semimembranosus
6	gluteus minimus	20	seventh lumbar nerve
7	ilium	21	spinal ganglion
8	inferior gluteal nerve	22	superior gluteal nerve
9	inferior hemorrhoidal nerve	23	tensor fasciae latae
10	lumbosacral cord	24	tibial nerve
11	muscular branch of sciatic nerve	25	transverse process of seventh lumbar vertebra
12	obturator nerve	26	transverse process of first caudal vertebra
13	posterior femoral cutaneous nerve	27	vastus lateralis
14	pudendal nerve	28	ventral ramus of sixth lumbar nerve

branch to the biceps, semitendinosus, and semimembranosus. It then passes across the adductor femoris and the semimembranosus. Above the knee it gives off the sural nerve, which passes to the lateral surface of the gastrocnemius, extending as far as the ankle. Near the knee the sciatic divides into the common peroneal and tibial nerves, which supply the muscles of the lower leg.

DISSECTION OF THIGH   Cut and remove the origins of the gluteus medius, the gluteus maximus, the caudofemoralis, and the biceps femoris. Remove the piriformis. Cut away the ilium and the dorsal aspect of the vertebral column, tracing the origins of the sciatic nerve as illustrated in Figure 72.

GLUTEAL NERVES   The superior gluteal nerve passes laterally around the dorsal side of the ilium, deep to the gluteus minimus, to supply the gluteus medius, gluteus minimus, and gemellus superior. The inferior gluteal nerve passes posteriorly, supplying the caudofemoralis and the gluteus maximus.

PUDENDAL NERVE   The pudendal nerve arises from the second and third sacral nerves. It passes lateral to the muscles of the tail and divides into two branches. One branch goes to the anus; the other goes to the penis (in the male) or to the urogenital sinus and adjacent structures (in the female).

POSTERIOR FEMORAL CUTANEOUS NERVE   The posterior femoral cutaneous nerve arises from the second and third sacral nerves and passes out of the pelvis with the inferior gluteal artery and vein. It gives branches to the anal region and to the biceps, and extends along the lateral aspect of the thigh as far as the popliteal space.

INFERIOR HEMORRHOIDAL NERVE   The inferior hemorrhoidal nerve arises from the second and third sacral nerves and passes across the lateral surface of the rectum to be distributed to the bladder and rectum.

COCCYGEAL NERVES   Seven or eight small coccygeal nerves exit from the intervertebral foramina of the first seven or eight caudal vertebrae. The dorsal rami innervate the muscles and skin on the dorsal side of the tail; the ventral rami innervate the muscles and skin on the ventral side of the tail.

SYMPATHETIC TRUNKS   The sympathetic trunks consist of a double line of ganglia and interconnecting nerves which lie on either side of the vertebral column from the base of the skull to the tail. The ganglia of the sympathetic trunks are connected to the spinal nerves by slender branches termed rami communicantes. The sympathetic trunks give off nerves to the thoracic and abdominal viscera, and these branches commonly pass through one or more of the autonomic nerve plexuses before reaching their destinations. See Figure 48 (p. 69) and Figure 70 (p. 97).

CERVICAL PORTION OF SYMPATHETIC TRUNKS   At the anterior end of the cervical portion of the sympathetic trunk is the superior cervical ganglion, which lies on the cranioventral side of the nodose ganglion of the vagus nerve. From the superior cervical ganglion the sympathetic trunk extends posteriorly, closely bound to the vagus. Near the first rib the sympathetic trunk separates from the vagus and enters the middle cervical ganglion (sometimes missing). From the middle cervical ganglion branches pass on either side of the subclavian artery to join the inferior cervical (stellate) ganglion, which lies between the first and second ribs. The inferior cervical ganglion gives communicating branches to the lower cervical and upper thoracic nerves. The middle and inferior cervical ganglia give off slender cardiac nerves which accompany the vagus to the heart, but these are too small to be identified easily.

THORACIC PORTION OF SYMPATHETIC TRUNKS

In the thoracic region the sympathetic trunk lies near the heads of the ribs. The first three thoracic sympathetic ganglia are fused to form the inferior cervical ganglion; below the third thoracic nerve there is a sympathetic ganglion corresponding to each spinal nerve. Observe the delicate rami communicantes which connect the ganglia and the spinal nerves.

The thoracic portion of the sympathetic trunk gives rise to the greater and lesser splanchnic nerves, which pierce the diaphragm to join the celiac and superior mesenteric plexuses and ganglia (see Figs. 42, p. 55, and 48, p. 69). The sympathetic trunk passes through the diaphragm and into the abdominal cavity dorsal to the splanchnic nerves.

ABDOMINAL PORTION OF SYMPATHETIC TRUNKS

In the abdomen the two trunks lie near the midline on the bodies of the lumbar vertebrae and are concealed by the iliopsoas muscles. They extend into the sacral region, becoming gradually smaller until they are lost. In the lumbar and sacral regions the sympathetic ganglia correspond to the spinal nerves and are connected with them by rami communicantes.

The above description covers only those portions of the sympathetic system that can be identified with relative ease in gross dissection. The autonomic nervous system includes numerous structures too minute for inclusion in a survey of gross anatomy.

THE EYE

Review the structure of the bony orbit.

NICTITATING MEMBRANE   Separate the upper and lower lids and observe the nictitating membrane, a prominent fold which originates at the medial corner of the eye and covers most of its surface.

CONJUNCTIVA   The conjunctiva is the mucous membrane of the eye. It covers the inner surface of the lids and both sides of the nictitating membrane, and is reflected over the forepart of the eyeball. Pull the lids away from the eye and observe the conjunctiva at the point where the lid and the eyeball meet.

DISSECTION OF ORBIT   Remove the zygomatic arch, the ramus of the dentary bone, and the muscles of the jaw as necessary to expose the contents of the orbit.

PERIORBITA AND GLANDS OF EYE   The eye and its muscles are enclosed in a tough, membranous sac termed the periorbita. Also enclosed by the periorbita are two glands: the lacrimal gland and the infraorbital gland. The infraorbital gland lies on the ventrolateral floor of the orbit; its duct opens into the mouth just posterior to the molar tooth. The lacrimal gland is a flat gland about one centimeter in diameter. It lies on the dorsolateral surface of the eyeball. The clear lacrimal fluid is conveyed to the eye by numerous minute ducts. It collects at the medial angle of the eye and drains into two small lacrimal canals which open near the medial corner of each lid. These canals unite to form the nasolacrimal duct which opens into the nasal cavity.

Remove the periorbita, the infraorbital gland, and the lacrimal gland to make a dissection similar to Figure 73.

MUSCLES OF EYE   The muscles of the eyelids are the orbicularis oculi and the levator palpebrae superioris. The orbicularis oculi lies in and around the eyelids. It is a sphincter which closes the lids. The levator palpebrae superioris is a slender muscle which originates near the optic foramen and passes dorsal to the eyeball between the superior rectus and the lacrimal gland to insert on the upper lid. It raises the lid.

Movements of the eye are effected by seven muscles which originate within the bony orbit and insert on the eyeball. The four rectus muscles originate around the optic foramen and are inserted by thin tendons which attach near the equator of the eyeball. They are named according to their positions: superior rectus,

FIG. 73.

LATERAL VIEW OF THE EYE

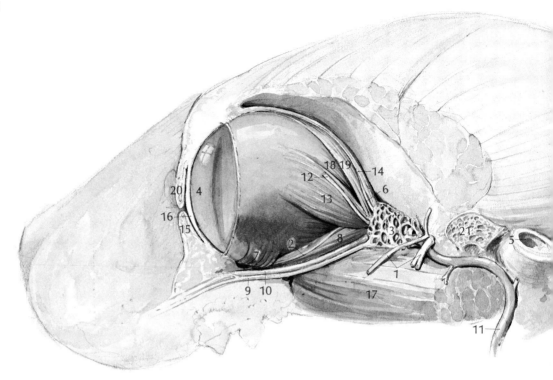

1 branches of mandibular nerve
2 branch of oculomotor nerve to inferior oblique muscle
3 carotid plexus
4 cornea
5 external auditory canal
6 frontal branch of ophthalmic nerve
7 inferior oblique muscle
8 inferior rectus muscle
9 infraorbital artery
10 infraorbital nerve
11 internal maxillary artery
12 lacrimal branch of maxillary nerve
13 lateral rectus muscle
14 levator palpebrae superioris
15 lower lid
16 nictitating membrane
17 pterygoid muscles
18 retractor bulbi muscle
19 superior rectus muscle
20 upper lid
21 zygomatic process of temporal bone, cut

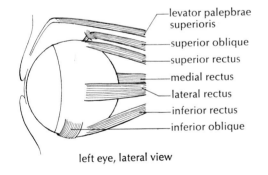

left eye, lateral view

inferior rectus, lateral rectus, and medial rectus.

The retractor bulbi is a cone-shaped muscle which originates around the optic foramen. It lies deep to the rectus muscles, inserts posterior to them, and is divided into four parts which alternate with the four recti.

The inferior oblique arises from the maxillary bone. It passes laterally on the ventral surface of the eyeball, crosses the inferior rectus, and inserts on the lateral surface of the eye near the insertion of the lateral rectus.

The superior oblique originates immediately above the optic foramen and passes anteriorly, dorsal to the medial rectus. Near the rim of the orbit it narrows to form a round tendon which passes through a fibrous band termed the trochlea, which acts as a pulley. After passing through the trochlea, the tendon of the superior oblique turns laterally, widens, and inserts near the insertion of the superior rectus.

REMOVAL AND SECTION OF EYE

Trim away the dorsal part of the orbit as necessary to expose the superior oblique and the medial rectus. Cut the muscles of the eye near their insertions, cut the optic nerve, and remove the eye. Make a horizontal cut around the eyeball and remove the dorsal half. Remove the other eye and cut it around the equator, dividing it into anterior and posterior halves.

EXTERNAL TUNIC

The eye is composed of three concentric layers or tunics. The external tunic consists of the sclera and cornea. The sclera, or "white" of the eye, is an opaque protective covering composed of white

FIG. 74.

DORSAL VIEW OF THE EYES

1   anterior chamber
2   ciliary body
3   choroid
4   cornea
5   cribiform plate of ethmoid
6   cut edge of conjunctiva
7   frontal branch of
    ophthalmic nerve
8   iris
9   lateral rectus muscle
10 lens
11 levator palpebrae
    superioris
12 medial rectus muscle
13 nasal conchae
14 optic chiasm
15 optic foramen
16 optic nerve
17 ora serrata
18 retina
19 retractor bulbi
20 sclera
21 superior oblique muscle
22 superior rectus muscle
23 suspensory ligament
    of lens
24 trochlea
25 vitreous humor

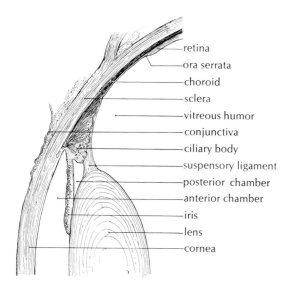

sagittal section of the human eye
(after Gray)

fibrous tissue and elastic fibers. It covers the posterior three
fourths of the eyeball and is continuous anteriorly with the
transparent cornea, which constitutes the exposed surface of the eye.

MIDDLE TUNIC The vascular middle tunic is composed of the iris, ciliary body,
and choroid. The choroid is a pigmented and richly vascular layer
lying between the sclera and the retina. It is continuous anteriorly
with the ciliary body. The ciliary body, which gives attachment
to the suspensory ligament of the lens, extends from the ora
serrata (anterior margin of the sensory part of the retina) to the
iris. It consists of a thickened portion of the vascular tunic
plus the ciliary muscle, a band of involuntary muscle fibers.
Its inner surface is covered by a very thin, nonsensory extension
of the retina. Observe the ciliary body as seen in the anterior half
of the eye which was cut around the equator, and note that its surface
is marked by numerous radial folds.

Anteriorly the ciliary body is continuous with the iris, a thin
circular sheet that lies between the lens and the cornea. The iris is
highly pigmented, and represents a fusion of the anterior part of
the vascular tunic and the anterior, nonsensory part of the retina.
Centrally the iris is pierced by an opening termed the pupil.
In bright light the pupil is a narrow vertical slit; in medium light it
becomes elliptical, and in a cat killed with chloroform it is
wide and circular.

INNER TUNIC The inner tunic of the eye is the retina, most of which is sensitive to
light. Posteriorly the retina is continuous with the optic nerve.
Anteriorly the photoreceptive part of the retina extends to the
ora serrata, just behind the ciliary body. From this point the retina
continues forward as a thin, nonsensory layer covering the
posterior portion of the iris and the ciliary body. The area of
most acute vision is the area centralis, which lies in the posterior part

of the retina near the optic nerve and in direct line with the optical axis of the eye. It is not grossly visible in the preserved eye.

TAPETUM LUCIDUM   Pull the retina away from the choroid and observe that within the choroid there is a lustrous oval area above the optic nerve. This is the tapetum lucidum, which is responsible for the eyeshine seen when car lights are reflected in a cat's eyes at night. It is an adaptation for vision in limited light.

LENS   The lens is suspended near the anterior part of the eye by the suspensory ligament of the lens, which is attached to the ciliary body. The curvature of the lens, and thus its focus, can be modified somewhat by the contraction of the ciliary muscle. This process is termed accommodation.

CAVITIES OF EYE   The cavities of the eye contain fluids termed the aqueous humor (anterior to the lens) and the vitreous humor (posterior to the lens). That portion of the eye anterior to the lens is divided by the iris into an anterior chamber (between the cornea and the iris) and a posterior chamber (between the iris and the suspensory ligament of the lens).

THE EAR

DISSECTION OF EAR   The dissection of the ear is complicated by the small size of the structures involved and by the fact that the inner ear is embedded in bone. The ear may, therefore, best be studied by reference to illustrations, models, and a demonstration dissection. If a student dissection of the ear is to be attempted, proceed as follows: remove the brain from one of the sagittal sections of the head. Cut through the head anterior to the external auditory canal and remove that portion of the skull anterior to the ear. Likewise, cut through the skull and remove the portion posterior to the ear. Soak the specimen for several days in 10 per cent nitric acid to decalcify the bone.

Using small bone clippers, approach the ear from the anterior aspect, chipping away bone as necessary to expose the structures of the external and middle ear as illustrated in Figure 75. Also see demonstration dissections in which these structures are exposed from the superior and posterior views. The inner ear is shown schematically in Figure 76; it is not feasible to dissect it out of the petrous part of the petromastoid bone as a unit. Its position can best be determined by making sections of the surrounding bone. This can be done with a razor blade if the bone has been thoroughly decalcified.

The ear consists of three parts: (1) the external ear, consisting of the pinna (cartilaginous portion outside the head) and the external auditory canal; (2) the middle ear, consisting of the tympanic membrane, the middle ear cavity within the temporal bone, and the auditory ossicles; (3) the inner ear, consisting of the cavity within the petrous temporal bone and the membranous labyrinth of sacs and ducts within this cavity.

EXTERNAL EAR   The pinna consists of a thin auricular cartilage which is covered by skin. Six muscles connect the various parts of the auricular cartilage with each other, and fifteen muscles attach the cartilage to the head. In man there are but three muscles connecting the auricular cartilage to the head.

The auricular cartilage continues medially as a cartilaginous tube termed the external auditory canal. At the medial end of this canal is the tympanic membrane, which covers the external auditory meatus of the temporal bone.

106

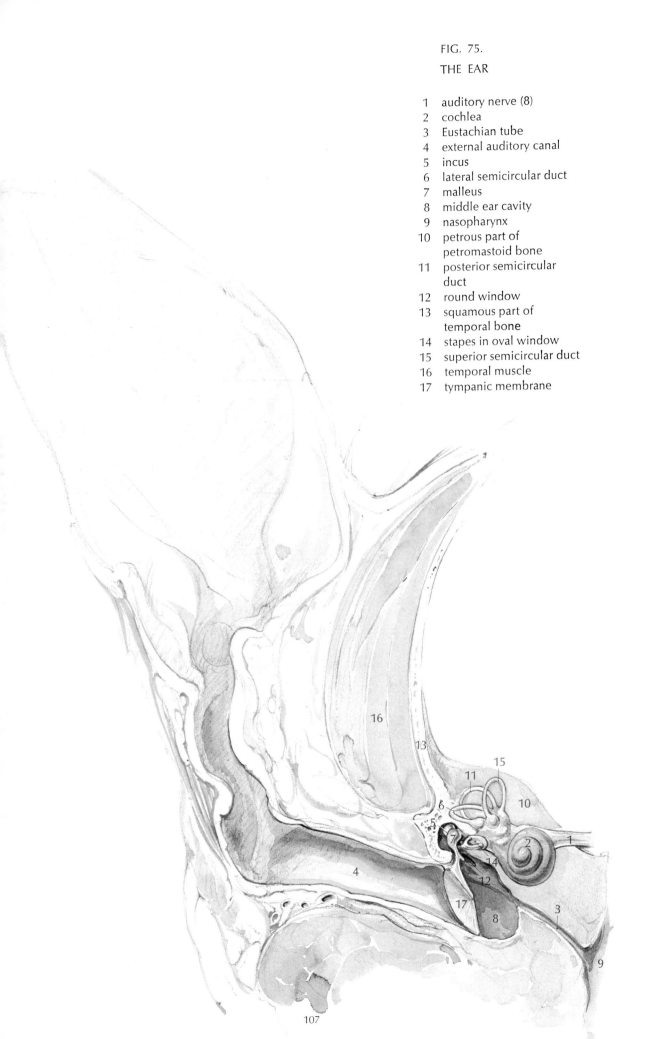

FIG. 75.

THE EAR

1 auditory nerve (8)
2 cochlea
3 Eustachian tube
4 external auditory canal
5 incus
6 lateral semicircular duct
7 malleus
8 middle ear cavity
9 nasopharynx
10 petrous part of
   petromastoid bone
11 posterior semicircular
   duct
12 round window
13 squamous part of
   temporal bone
14 stapes in oval window
15 superior semicircular duct
16 temporal muscle
17 tympanic membrane

FIG. 76.

THE RIGHT MEMBRANOUS
LABYRINTH OF THE HUMAN EAR,
MEDIAL VIEW (after Sobotta)

1 ampulla
2 cochlea
3 cochlear nerve
4 lateral semicircular duct
5 posterior semicircular
   duct
6 sacculus
7 superior semicircular
   duct
8 utriculus
9 vestibular nerve

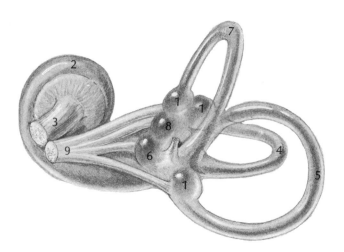

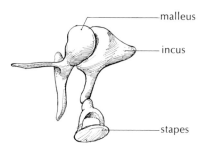

human middle ear ossicles
(after Sobotta)

MIDDLE EAR Vibrations are conveyed from the tympanic membrane to the inner ear by an articulated chain of three minute bones termed auditory ossicles: the malleus (hammer), incus (anvil), and stapes (stirrup). The malleus is attached to the tympanic membrane. The incus lies between the malleus and the stapes, which fits into an opening termed the oval window in the medial wall of the middle ear cavity.

Ventral and posterior to the oval window is a similar opening termed the round window. It is covered by a delicate membrane. Examine a skull in which the lateral portion of the middle ear cavity has been removed to expose the medial wall, and identify the round and oval windows. In the dry skull the round and oval windows form a direct connection between the cavity of the inner ear and the cavity of the middle ear.

The Eustachian tube is a partially cartilaginous tube about two centimeters long which extends between the nasopharynx and the ventromedial wall of the middle ear cavity. See its opening into the nasopharynx as illustrated in Figure 30, page 39.

INNER EAR The inner ear consists of the bony labyrinth, which is a cavity within the petrous part of the petromastoid bone, and the membranous labyrinth, which is a delicate membranous system of ducts and sacs. The membranous labyrinth lies within the bony labyrinth and closely resembles it in shape. The space between the membranous labyrinth and the bony labyrinth is filled with a clear, lymph-like fluid, the perilymph. The membranous labyrinth itself is filled with a similar fluid, the endolymph. The ramifications of the auditory nerve (cranial nerve number 8) terminate in sensory areas in the membranous labyrinth.

The membranous labyrinth consists of a spiral part, the cochlea, two membranous sacs termed the utriculus and the sacculus, and three delicate curved tubes termed the lateral, posterior, and anterior semicircular ducts.

BIBLIOGRAPHY

Crouch, James E. *Text-Atlas of Cat Anatomy*. Philadelphia: Lea & Febiger, 1969.

Gray, Henry. *Anatomy of the Human Body,* ed. Charles M. Goss. 26th ed. Philadelphia: Lea and Febiger, 1954.

Hyman, Libbie H. *Comparative Vertebrate Anatomy.* 2nd ed. Chicago: University of Chicago Press, 1942.

Mivart, St. George J. *The Cat*. New York: Charles Scribner's Sons, 1881.

Reighard, Jacob, and H. S. Jennings. *Anatomy of the Cat,* ed. Rush Elliott. 3rd ed. New York: Holt, Rinehart and Winston, 1935.

Taylor, William T., and Richard J. Weber. *Functional Mammalian Anatomy (with Special Reference to the Cat)*. Princeton, N.J.: D. Van Nostrand, 1951.

DEFINITIONS OF DESCRIPTIVE TERMS

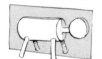

Median sagittal plane

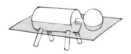

Horizontal or frontal plane

ansverse or cross plane

Right and *left* are determined with reference to the orientation of the specimen, not with reference to the orientation of the observer.

Dorsal: toward the back, or upper side.

Ventral: toward the abdomen, or under side.

Lateral: pertaining to the side of the body.

Medial: pertaining to the middle, or midline of the body.

Distal: pertaining to a position removed from the center of the body or from the origin of a structure.

Proximal: pertaining to a position close to the center of the body or to the origin of a structure.

Anterior, cephalic, or *cranial:* toward the head.

Posterior or *caudal:* toward the tail.

Deep or *central:* near the middle of the trunk or of a limb.

Superficial: near the surface of the trunk of or a limb.

Median sagittal plane: the plane which divides the body into right and left halves.

Sagittal plane: any plane parallel to the median sagittal plane.

Horizontal or *frontal plane:* at right angles to the sagittal plane and parallel to the dorsal and ventral surfaces.

Transverse or *cross plane* : at right angles to both the sagittal and horizontal planes.

Superior: above.

Inferior: below.

INDEX

(Light-face numbers indicate pages; bold-face numbers indicate figures)

mesovarium, 58; **42**
metacarpals, **1**, **11**
metacromion, **10**
metatarsals, **1**, **13**
metencephalon, 79; **60**
middle cerebellar peduncle, 79; **67**, **68**
    cerebral artery, **62**
    cervical ganglion, 100
    colic artery, 47; **36**
    ear, 106; **75**
    hemorrhoidal artery, 70; **44**, **49**
    lobe of lung, 41; **31**
    meningeal, 66
    tunic of eye, 104; **74**
molar gland, 37; **29**
    tooth, 38; **2**, **3**, **6**
Monro, foramen of, 76, 84; **55**, **63**
motor root of spinal nerve, 92
movable joints, 3
multangular bone, greater, **11**
        lesser, **11**
MUSCLES
    acromiodeltoid, 21; **16**, **17**
    acromiotrapezius, 23; **16**, **17**
    actions and attachments of, 16
    adductor femoris, 34; **25**, **26**, **27**, **28**
        longus, 34; **25**, **27**, **28**
    anconeus, 26; **20**
    biceps brachii, 26; **19**, **21**
    brachialis, 26; **19**, **20**
    brachioradialis, 23; **14**, **17**
    caudofemoralis, 33; **24**, **26**
    ciliary, 104
    clavobrachialis, 21; **14**, **16**
    clavotrapezius, 21; **15**, **16**
    cleidomastoid, 24; **18**
    coracobrachialis, 26; **19**, **21**
    cutaneous maximus, 17
    digastric, 20; **15**
    epitrochlearis, 19; **14**
    extensor carpi obliquus, 26; **20**
        radialis brevis, 23; **14**, **17**, **20**
           longus, 23; **14**, **17**, **20**
        ulnaris, 23; **17**, **20**
        digitorum communis, 23; **17**
           lateralis, 23; **17**
           longus, 30; **24**, **26**
        of first and second digits, 26; **20**
    external intercoastal, 29; **23**
        oblique, 35; **22**
    flexor carpi radialis, 19; **14**
        ulnaris, 19; **14**
        digitorum longus, 31; **25**, **27**
           profundus, 27; **21**
           sublimis, 26; **21**
        hallucis longus, 31; **25**, **27**
    gastrocnemius, 30; **24**, **25**, **26**, **27**
    gemellus superior, **71**, **72**
    genioglossus, 39; **30**
    geniohyoid, **18**, **22**
    gluteus maximus, 33; **24**, **26**
        medius, 33; **26**
        minimus, **71**, **72**
    gracilis, 30; **25**
    hyoglossus, 39
    iliopsoas, 30; **25**, **28**, **39**
    inferior oblique, 103; **73**
        rectus, 102, 103; **73**
    infraspinatus, 26; **20**
    intercostals, 29
    internal oblique, 36
    intertransversus, **70**

(Muscles, continued)
    lateral rectus, 102, 103; **73**, **74**
    latissimus dorsi, 23; **16**, **17**
    levator ani, **49**
        palpebrae superioris, 102; **73**, **74**
        scapulae ventralis, 23; **16**, **17**
    longus capitis, **23**
    masseter, 20; **15**, **16**
    medial rectus, 102, 103; **74**
    mylohyoid, 20; **15**
    oblique of eye, 103; **73**, **74**
        of abdominal wall, 35, 36
    obturator internus, **71**
    orbicularis oculi, 102
    palmaris longus, 19; **14**
    papillary, 74; **52**, **53**
    pectineus, 31; **25**, **27**, **28**
    pectoantebrachialis, 19; **14**
    pectoralis major, 19; **14**
        minor, 19; **14**
    peroneus brevis, 30; **24**, **26**
        longus, 30; **24**, **26**
        tertius, 30; **24**
    piriformis, **71**
    plantaris, 35; **27**
    platysma, 17
    pronator quadratus, 27
        teres, 19; **14**, **21**
    psoas minor, 39, **42**
    pterygoid, **73**
    quadratus femoris, **72**
    quadriceps femoris, 34, 35; **26**, **27**, **28**
    rectus abdominis, 36; **19**, **25**
        femoris, 35; **27**, **28**
    recti of eye, 102, 103; **73**, **74**
    retractor bulbi, 103; **73**, **74**
    rhomboideus, 25; **20**, **22**
        capitis, 25; **20**, **22**
    sacrospinalis, 29
    sartorius, 30; **24**, **25**
    scalenes, 28; **22**, **23**
    semimembranosus, 32; **26**, **27**
    semitendinosus, 32; **24**, **25**, **27**
    serratus dorsalis caudalis, 28, 29; **23**
        cranialis, 28, 29; **20**, **22**, **23**
        ventralis, 24; **19**, **20**, **22**
    soleus, 30; **24**, **26**
    spinodeltoid, 21; **16**, **17**
    spinotrapezius, 23; **16**, **17**
    splenius, 28; **20**, **22**
    sternohyoid, 24; **15**, **18**, **22**
    sternomastoid, 20; **15**, **16**
    sternothyroid, 24; **18**, **22**
    subscapularis, 26; **21**
    superior oblique, 103; **73**, **74**
        rectus, 102; **73**, **74**
    supinator, 26; **20**
    supraspinatus, 25; **20**, **21**
    temporal, 20; **15**, **16**
    tensor fasciae latae, 30; **24**, **26**, **27**
    tenuissimus, 32; **26**
    teres major, 26; **20**, **21**
        minor, 26; **20**
    thyrohyoid, **18**, **22**
    tibialis anterior, 30; **24**, **25**, **27**
        posterior, 35; **25**, **27**
    transversus abdominis, 36
        costarum, 28; **19**, **22**, **23**
    triceps brachii, 23; **16**, **17**, **18**
    vastus intermedius, 35
        lateralis, 34, 35; **26**, **28**
        medialis, 34, 35; **27**, **28**

(Muscles, continued)
    xiphihumeralis, 19; **14**
    musculi pectinati, 73; **52**
    musculocutaneous nerve, 94; **69**
    myelencephalon, 79; **60**
    mylohyoid muscle, 20; **15**

nares, external, **2**
    internal, 5, 40; **6**, **30**
nasal bone, 2, 4, 5
    conchae, 3, 7; **5**, **30**
nasolacrimal canal, 4
    duct, 102
nasopalatine duct, **30**
navicular bone, **13**
neck of femur, **12**
    of rib, **9**
neopallium, 77; **57**
nephron, 54
NERVES
    abducens, 82, 86; **8**, **62**, **64**, **68**
    anterior ventral thoracic, 94; **69**
    auditory, 82, 86; **8**, **62**, **64**, **68**, **75**
    axillary, 94; **45**, **46**, **69**
    cardiac, 100
    cerebrospinal, defined, 75
        described, 85-88, 92-100
    cervical, 93; **45**, **46**, **69**
    coccygeal, 100; **72**
    cochlear, 76
    common peroneal, 100; **71**, **72**
    cranial, 85, 86, 88; **62**, **64**
    dorsal division of vagus, 88; **48**
    facial, 86; **62**, **64**
    femoral, 96; **50**, **70**
        cutaneous, posterior, 98, 100; **72**
    first subscapular, 94; **69**
        thoracic, 94; **69**
    genitofemoral, 96; **50**, **70**
    glossopharyngeal, 82, 86; **8**, **62**, **64**
    gluteal, inferior, 100; **71**, **72**
        superior, 100; **71**, **72**
    greater splanchnic, 101; **42**, **48**
    hypoglossal nerve, 88; **8**, **62**, **64**
    inferior gluteal, 100; **71**, **72**
        hemorrhoidal, 100; **72**
    infraorbital, **73**
    intercostal, 94, 96; **48**
    laryngeal, recurrent, 88; **48**
    lateral cutaneous, 96; **70**
    lesser splanchnic, 101; **42**, **48**
    long thoracic, 94; **69**
    lumbar, 96; **70**, **72**
    mandibular, 86; **64**
    maxillary, 86; **64**
    medial cutaneous, 94; **69**
    median, 94; **45**, **46**, **69**
    obturator, 96; **49**, **70**
    oculomotor, 82, 85, 86; **8**, **62**, **64**
    olfactory, 81, 85; **8**, **62**, **64**
    ophthalmic, 86; **8**, **64**
    optic, 85; **8**, **62**, **64**, **74**
    peroneal, common, 100; **71**, **72**
    phrenic, 93; **45**, **46**, **48**, **69**
    posterior femoral cutaneous, 98, 100; **72**
        ventral thoracic, 94; **69**
    pudendal, 98, 100; **72**
    radial, 94; **45**, **46**, **69**
    recurrent laryngeal, 88; **48**
    sacral, 96, 98, 99, 100; **70**, **72**
    saphenous, 96; **50**, **70**
    sciatic, 98, 99; **71**, **72**